MOUNT RAINIER

Active Cascade Volcano

MOUNT RAINIER

Active Cascade Volcano

Research Strategies for Mitigating Risk from a High, Snow-Clad Volcano in a Populous Region

U.S. Geodynamics Committee
Board on Earth Sciences and Resources
Commission on Geosciences, Environment, and Resources
National Research Council

NATIONAL ACADEMY PRESS
Washington, D.C. 1994

NOTICE: The project that is the subject of this report was approved by the Governing Board of the National Research Council, whose members are drawn from the councils of the National Academy of Sciences, the National Academy of Engineering, and the Institute of Medicine. The members of the committee responsible for the report were chosen for their special competences and with regard for appropriate balance.

This report has been reviewed by a group other than the authors according to procedures approved by a Report Review Committee consisting of members of the National Academy of Sciences, the National Academy of Engineering, and the Institute of Medicine.

Support for this study was provided by the U.S. Department of Energy, National Aeronautics and Space Administration, National Science Foundation, and U.S. Geological Survey.

Library of Congress Catalog Card No. 94-66300
International Standard Book Number 0-309-05083-9

Copies of this report are available from:

National Academy Press
2101 Constitution Avenue, N.W.
Washington, D.C. 20418

800-624-6242
202-334-3313 (in the Washington Metropolitan Area)

Cover photograph of Mount Rainier courtesy of Richard S. Fiske, Smithsonian Institution. Frontispiece lithograph of Mount Rainier is from the expedition of Captain George Vancouver to northwestern America in 1790–1795. The lithograph appeared in the three-volume work *A Voyage of Discovery to the North Pacific Ocean and Round the World*, which was published in 1798 by G. G. and J. Robinson, Paternoster-Row, and J. Edwards, Pall-Mall. Courtesy of the Special Collections Division, University of Washington Libraries (negative number 8184).

Printed in the United States of America

U.S. GEODYNAMICS COMMITTEE

ROBIN BRETT, U.S. Geological Survey, *Chair*
DON L. ANDERSON, California Institute of Technology
KEVIN T. BIDDLE, Exxon Exploration Company
MARK P. CLOOS, University of Texas at Austin
WILLIAM DICKINSON, University of Arizona
RICHARD S. FISKE, Smithsonian Institution
RAYMOND JEANLOZ, University of California, Berkeley
KENNETH LARNER, Colorado School of Mines
ELIZABETH MILLER, Stanford University
LYNN M. WALTER, University of Michigan
ROBERT S. YEATS, Oregon State University

Former Members Whose Terms Expired During the Reporting Period

T. MARK HARRISON, University of California, Los Angeles
WILLIAM J. HINZE, Purdue University
JOHN C. MUTTER, Lamont-Doherty Earth Observatory
ROBERT A. PHINNEY, Princeton University
SIGMUND SNELSON, Shell Oil Company

Workshop Organizers/Working Group Report Editors

THOMAS J. CASADEVALL, U.S. Geological Survey
STEPHEN D. MALONE, University of Washington
DONALD A. SWANSON, U.S. Geological Survey

National Research Council Staff

KEVIN D. CROWLEY, Program Officer
BRUCE B. HANSHAW, Program Officer
SHELLEY A. MYERS, Project Assistant

COMMISSION ON GEOSCIENCES, ENVIRONMENT, AND RESOURCES

The National Academy of Sciences is a private, nonprofit, self-perpetuating society of distinguished scholars engaged in scientific and engineering research, dedicated to the furtherance of science and technology and to their use for the general welfare. Upon the authority of the charter granted to it by the Congress in 1863, the Academy has a mandate that requires it to advise the federal government on scientific and technical matters. Dr. Bruce Alberts is president of the National Academy of Sciences.

The National Academy of Engineering was established in 1964, under the charter of the National Academy of Sciences, as a parallel organization of outstanding engineers. It is autonomous in its administration and in the selection of its members, sharing with the National Academy of Sciences the responsibility for advising the federal government. The National Academy of Engineering also sponsors engineering programs aimed at meeting national needs, encourages education and research, and recognizes the superior achievements of engineers. Dr. Robert M. White is president of the National Academy of Engineering.

The Institute of Medicine was established in 1970 by the National Academy of Sciences to secure the services of eminent members of appropriate professions in the examination of policy matters pertaining to the health of the public. The Institute acts under the responsibility given to the National Academy of Sciences by its congressional charter to be an adviser to the federal government and, upon its own initiative, to identify issues of medical care, research, and education. Dr. Kenneth I. Shine is president of the Institute of Medicine.

The National Research Council was organized by the National Academy of Sciences in 1916 to associate the broad community of science and technology with the Academy's purposes of furthering knowledge and advising the federal government. Functioning in accordance with general policies determined by the Academy, the Council has become the principal operating agency of both the National Academy of Sciences and the National Academy of Engineering in providing services to the government, the public, and the scientific and engineering communities. The Council is administered jointly by both Academies and the Institute of Medicine. Dr. Bruce Alberts and Dr. Robert M. White are chairman and vice-chairman, respectively, of the National Research Council.

ACKNOWLEDGMENTS

The U.S. Geodynamics Committee wishes to acknowledge the assistance of the many scientists who made this report possible. Tom Casadevall, Steve Malone, Barbara Samora, and Don Swanson published two articles in EOS in 1992 and 1993 that focused the interest of the volcanological community on Mount Rainier[1]. The committee used the information in these articles to develop this report. At the committee-sponsored workshop in 1992, Russell Blong, John Delaney, John Dvorak, Al Eggers, Peter Frenzen, Gary Machlis, Peter May, Patrick Pringle, Barbara Samora, Dal Stanley, and Ed Wolfe developed the working group reports that the committee used to prepare the present report. The working group reports were collated and edited by Tom Casadevall, Steve Malone, and Don Swanson, who also planned, organized, and ran the workshop on behalf of the committee.

The committee is pleased to acknowledge the University of Washington and U.S. Geological Survey for their support of the workshop. The Committee also thanks the U.S. Department of Energy, National Aeronautics and Space Administration, National Science Foundation, and U.S. Geological Survey, whose continuing support of the committee made this report possible.

---

[1] Swanson, D. A., Malone, S. D., and Samora, B. A., 1992, Mount Rainier: a Decade Volcano: EOS, v. 73, p. 177, 185-186; Swanson, D. A., Malone, S. D., and Casadevall, T., 1993, Mitigating the hazards of Mount Rainier: EOS, v. 74 (12), p. 133.

PREFACE

The United Nations designated the 1990s as the International Decade for Natural Disaster Reduction (IDNDR; United Nations Resolution 42/169/1987). The objective of the IDNDR is to reduce threats to human life and development from natural hazards through the application of science and technology. The Science and Technical Committee of IDNDR has endorsed the concept of Demonstration Projects—focused scientific studies of specific natural hazards such as volcanoes—to meet this objective.

The designation of Volcano Demonstration Projects for the IDNDR is being coordinated by the International Association of Volcanology and Chemistry of the Earth's Interior (IAVCEI), one of the seven associations comprising the International Union of Geodesy and Geophysics. To date, IAVCEI has designated 14 Decade Volcanoes, including Mount Rainier, for focused study. These volcanoes represent a variety of eruptive styles and potential hazards. They are generally located in accessible, populated regions and are geologically active but not well studied. Scientific research on these volcanoes is likely to improve the understanding of potential hazards in similar environments worldwide.

Mount Rainier was selected as a Decade Volcano for several reasons. It has an extensive but poorly studied geological and historical record of activity, including lava flows, ash eruptions, avalanches, and mudflows. The volcano thus poses a hazard to surrounding, highly populated regions, particularly the Seattle-Tacoma metropolitan area. It poses an additional hazard because of its extensive cover of snow and ice, which, if melted rapidly, could produce catastrophic floods and mudflows. Study of the volcano as a Decade Volcano Demonstration Project is likely to improve the understanding of these hazards and, concomitantly, to reduce risks to life and property in the region.

As a first step in developing a Volcano Demonstration Project for Mount Rainier, the U.S. Geodynamics Committee sponsored a workshop to draft a research plan for the volcano. A three-day workshop was held at the University of Washington, Seattle, on September 18-20, 1992, and involved about 75 earth scientists, experts in natural hazards and mitigation, and representatives of government agencies. The first two days of the workshop were devoted to a review of the geology and geophysical setting of Mount Rainier and surrounding regions and included a field trip to the volcano. On the third day of the workshop, participants formed six working groups to draft a science plan for the volcano. This working group document was edited by the workshop organizers and was used by the U.S. Geodynamics Committee to prepare the present report.

This report presents a science plan for the study of Mount Rainier as a Decade Volcano Demonstration Project and addresses the application of scientific results to the assessment of volcanic hazards and mitigation of risk. Although the science plan focuses primarily on research needed to understand the development and behavior of the volcano and to monitor potential hazards, the committee recognizes that scientific research alone will not advance the goals of the IDNDR program to mitigate risk from volcanic hazards. Accordingly, this report also addresses issues of communication and coordination among geoscientists, social scientists, planners, and responsible authorities, so that the results of this research can be used to support hazard reduction efforts. This link between research and application is an essential element of the IDNDR program.

The present report reflects many of the ideas of the workshop participants and organizers. However, the U.S. Geodynamics Committee accepts all responsibility for the report's content and recommendations.

CONTENTS

1
EXECUTIVE SUMMARY

Introduction

Mount Rainier is one of about two dozen active or recently active volcanoes in the Cascade Range, an arc of volcanoes in the northwestern United States and Canada. The volcano is located about 35 kilometers (km) southeast of the Seattle-Tacoma (Washington) metropolitan area, which has a population of more than 2.5 million. This metropolitan area is the high-technology industrial center of the Pacific Northwest and one of the commercial aircraft manufacturing centers of the United States. The rivers draining the volcano empty into Puget Sound, which has two major shipping ports, and into the Columbia River, a major shipping lane and home to approximately a million people in southwestern Washington and northwestern Oregon.

Mount Rainier is an active volcano. It last erupted approximately 150 years ago, and numerous large floods and debris flows have been generated on its slopes during this century. More than 100,000 people live on the extensive mudflow deposits that have filled the rivers and valleys draining the volcano during the past 10,000 years. A major volcanic eruption or debris flow could kill thousands of residents and cripple the economy of the Pacific Northwest. Despite the potential for such danger, Mount Rainier has received little study. Most of the geologic work on Mount Rainier was done more than two decades ago. Fundamental topics such as the development, history, and stability of the volcano are poorly understood.

Studies of the geologic history of Mount Rainier and other Cascade volcanoes suggest that major volcanic hazards, volcano-related events that pose threats to persons or property, are likely to include the following:

- *Volcanic eruptions.* The eruption of lava flows and tephra (particulate materials such as ash).
- *Edifice failure.* The gravitational collapse of a portion of the volcano.
- *Glacier outburst floods (jökulhlaups).* The sudden release of meltwater from glaciers and snowpack or from glacier-dammed lakes on the edifice.
- *Lahars, or debris flows, and debris avalanches.* Gravitational movement of commonly water-saturated volcanic debris down the steep slopes of the volcano and into nearby valleys.

Mount Rainier is capable of eruptions of small to very large magnitude, as measured by the Volcanic Explosivity Index of 4 to 5 that has been tentatively assigned to the explosive eruption that occurred between 30,000 and 100,000 years ago. Based on past activity, the most likely future eruptive event at Mount Rainier is the extrusion of a lava flow at the summit, possibly accompanied by tephra eruptions. Lava flows would likely be restricted to valley floors within or a short distance outside of Mount Rainier National Park, where they would destroy roads, buildings, and other fixed installations. The sluggish motion of these flows would probably permit people to evacuate safely from areas at risk, so little loss of life would be expected. However, steam columns and nighttime reflections of the glowing surfaces of lava flows from clouds could be visible from the Seattle-Tacoma metropolitan area, possibly creating an unwarranted sense of impending crisis.

Explosive eruptions from Mount Rainier could send clouds of tephra high into the atmosphere where they would be carried laterally by prevailing winds before settling to the ground. This tephra could be a major hazard to crops and other vegetation, poorly built structures, and machinery. The prevailing winds in western Washington are from southwest to northeast, so tephra from Mount Rainier would normally be carried away from the Seattle-Tacoma metropolitan area. Less frequently, winds blow from east to west, and at such times tephra could be scattered over much of Puget Lowland. This would disrupt commerce, travel (including flights at Seattle-Tacoma International Airport), and the daily lives of hundreds of thousands of people.

Major edifice failures, glacier outburst floods, and lahars could occur in the absence of volcanic eruptions because of the inherent instability of the volcanic edifice. Mount Rainier is a high volcano (4,392-meters above sea level with approximately 3,000 m of relief) that contains about 140 cubic kilometers (km^3) of structurally weak and locally altered rock capped by about 4.4 km^3 of snow and ice, all of which stand near the angle of repose. Ground shaking during an earthquake, or ground deformation due to intrusion of magma into the edifice, could cause the gravitational failure of a large sector of the volcano, producing catastrophic avalanches and debris flows and possibly triggering an eruption. Glacier outburst floods and lahars can also occur during heavy rainfalls or transient heating events that melt the snow and ice cover on the volcano.

Damage caused by debris flows could be substantial. Geologic mapping of surficial deposits in Mount Rainier National Park has shown that numerous debris flows have entered the rivers draining the volcano over the past several thousand years. The largest known debris flow from Mount Rainier, the Osceola Mudflow, traveled down the White River drainage system a distance of approximately 110 km about 4,500 to 5,000 years ago and transported at least 3 km^3 of rock debris, burying parts of the Puget Lowland that are now heavily populated. In the past 45 years, about two dozen debris flows and outburst floods have occurred at Mount Rainier, the majority in the Tahoma Creek-Nisqually River drainage. These debris flows traveled downstream as far as 16 km from their origin on the volcanic edifice.

Coordinated research that involves both geoscientists and social scientists should be undertaken to determine potential magnitudes and frequencies of potential hazards, their human and economic impacts, and strategies for using such information effectively to mitigate risk as part of this Decade Volcano Demonstration Project. A plan to achieve these objectives is outlined in this report.

Research

Regional studies are needed to address the formation and development of Mount Rainier within the Cascade volcanic arc environment. Of particular importance in this context are studies that address the following:

- tectonic processes that control the locations of volcanic vents;
- regional stress fields and their effects on volcanism, faulting, and seismicity;
- the crustal deformation field caused by magma injection, subduction, and glacial loading; and
- ages, distributions, and characteristics of tephras, lavas, and lahars.

Studies of the Mount Rainier edifice are also needed to address the development of the volcano in order to predict its future behavior. Of particular importance are studies that address these topics:

- the structure of the volcanic edifice and underlying crust;
- the history of edifice growth and failure;
- the geometry of hydrothermal and groundwater systems; and
- distributions of hydrothermally altered rocks.

A high degree of feedback between local and regional studies and between individual projects and investigators should be employed as part of the strategy for this Decade Volcano Demonstration Project. This project should be coordinated with ongoing research programs of federal, state, and academic scientists and should include the following elements:

1. Geologic mapping. Mapping the spatial and temporal distributions of eruptive and intrusive rocks, faults, hydrothermal alteration zones, surficial deposits, springs, fumaroles (vents that emit steam and other gases), and glaciers should be undertaken as part of the effort to understand the development of the Cascade volcanic arc and Mount Rainier edifice.

2. Petrologic and geochemical studies. Petrological and geochemical studies of Tertiary and Quaternary (particularly Holocene) rocks should be undertaken to address the physical characteristics and evolution of the magma system through time, to help establish stratigraphic relations among eruptive products, and to provide the basis for reconstructing patterns of hydrothermal alteration.

3. Geophysical surveys. Geophysical surveys should be undertaken to elucidate the structure of the volcanic edifice and underlying crust, including distributions of magma, intrusive bodies, faults, hydrothermal and groundwater systems, and glacier ice.

4. Lahar studies. Detailed mapping, including mapping of buried lahars, should be carried out to reconstruct the spatial and temporal distributions of these flows and to obtain volumetric estimates for each flow event.

5. Edifice stability assessment. Research on edifice stability should focus on mapping the distributions of hydrothermally altered rocks, faults, and dikes, which are mechanically weak and prone to failure. Research should also focus on the delineation of the hydrothermal system and the process of wallrock alteration, particularly beneath the glaciers that cover the edifice.

Volcano Monitoring

A program of volcano monitoring should be established at Mount Rainier to identify anomalous activity that could serve as an early warning of the occurrence of volcanic hazards such as eruptions, edifice collapse, and lahars. This monitoring program should include plans for the collection of adequate baseline data to provide a background of values with which to contrast anomalous behavior.

Monitoring should involve the following techniques, which have been developed and tested over the past several decades at active volcanoes around the world:

1. Seismic monitoring, using the present network of seismometers, to detect the movement of magma, glaciers, and rock on or beneath the volcano. The network should be upgraded with a minimum of two, three-component instruments to allow for the precise location of events on the edifice.

2. Ground-deformation monitoring, to detect edifice creep or the underground movement of magma. To this end, the present network of geodetic stations should be expanded with additional stations established at higher elevations on the edifice, and this local network should be integrated into the regional network. These local and regional networks should be monitored using real-time, continuous Global Positioning System (GPS) or they should be resurveyed using GPS at frequent intervals.

3. Monitoring hydrothermal activity, to detect changes in the composition or rates of emission of gases and fluids from the edifice. A program of fluid and gas sampling should be initiated to monitor the hydrothermal system on the volcanic edifice.

4. Monitoring changes in surface appearance, to detect changes in the snow and ice cover on the volcano. This monitoring should include visual observation, photogrammetry, infrared heat emission, and radar imagery.

5. Stream monitoring, to detect floods and lahars after they have formed and are moving downslope toward populated areas. To this end, a network of sensors tied into the existing seismic network should be installed in the major drainages on the volcanic edifice to detect the formation and movement of lahars.

Mitigation

Communities in the region must seek ways to reduce or mitigate risk to life and property from volcanic hazards while maintaining the strong economic base that derives in part from the desire of people to live, work, and play around the volcano. Effective risk mitigation can be successfully executed only within the context of a comprehensive strategy to understand the volcano. The success of mitigation efforts requires that the hazards themselves are well understood through a program of coordinated research as outlined in this report; that they can be recognized through effective monitoring before they reach a critical level; that warning of their occurrence can be communicated clearly, accurately, and quickly to public officials; and that public officials can and will act to put the appropriate risk-mitigating measures into operation.

The important elements of an effective mitigation program are these:

1. Communication is essential among the many groups that live and work around the volcano:

- *Within the scientific community*, to coordinate and disseminate research on the volcano. To this end, a Mount Rainier Hazards Information Network should be established on the Internet to disseminate past, current, and planned research and information on mitigation measures.
- *Between scientists and responsible authorities*, to provide precrisis information about volcanic hazards, and warnings of impending hazards. An emergency-response plan should be developed so that scientists involved in monitoring can provide responsible authorities with accurate and timely warnings of impending hazards and can keep officials informed during such events.
- *Between scientists and the public*, to inform the general public about the nature of volcanic hazards, people and property at risk, and options for risk reduction. Scientists should work with educators and National Park Service staff to develop and distribute high-impact educational materials, to provide presentations at schools and public meetings, and to develop displays on volcanic hazards and emergency response for visitors to Mount Rainier National Park.
- *Between responsible authorities and the public*, to communicate timely and accurate information and warnings about volcanic hazards to the public. Authorities should develop plans for such

communication and test those plans in simulated precrisis and crisis situations.

2. Planning and implementation of risk-mitigation measures should involve scientists, government, business, and citizens and should be coordinated and, where appropriate, integrated with other planning activities in the region. Several measures, including the following, should be considered for implementation in order to significantly reduce risk from volcanic hazards to people and property:

- analyses to identify regions and populations at risk;
- land use planning and economic incentives to discourage inappropriate use of high-risk areas; and
- engineering solutions to mitigate risks, where possible, from specific volcanic hazards.

An important contribution of geoscientists in these efforts should be the identification of areas at risk through the development of hazard maps, which are spatial representations of risk from hazards such as lava flows and debris flows. Geoscientists should work cooperatively with planners, engineers, social scientists, and legal professionals to ensure that these hazard maps contain appropriate data, presented in usable formats for risk-mitigation efforts.

Traditionally, natural scientists participate in mitigation efforts up to the point of public debate, providing information about hazards and, occasionally, about their effects on people. Social scientists, on the other hand, rarely become involved in the early stages of hazard studies. The U.S. Geodynamics Committee believes that more effective strategies could be developed and implemented if both groups work together, starting with geologic investigations and continuing throughout debate and implementation of mitigation measures. Similarly, social scientists, geoscientists, planners, engineers, citizens, and decision makers should work together, from hazard assessment through implementation, if the populations around Mount Rainier are to coexist in reasonable safety with the volcano.

Implementation

Implementation of the Mount Rainier Decade Volcano Demonstration Project is the responsibility of the scientific community, which should develop a plan to carry the project forward. This implementation plan should provide guidance on:

- priorities for research and monitoring activities based on scientific significance and value to risk-mitigation efforts;
- funding for research and monitoring activities deemed to be of high priority;
- mechanisms for coordinating the efforts of scientists to avoid unnecessary duplication, particularly in the use of instrumentation or collection of samples from wilderness and other environmentally sensitive areas with limited access; and
- mechanisms for balancing the needs of scientists for access, samples, and data with the needs of federal and state agencies to fulfill their research, public safety, and land-management missions.

To be effective, monitoring efforts will require continuity in funding, management, personnel, and facilities that can best be provided by federal and state agencies with responsibilities for volcano and hazards research. Nongovernment scientists should be encouraged to participate in monitoring activities in both data collection and advisory capacities, and the scientific community should have free and immediate access to monitoring data.

Many of the research, monitoring, and mitigation activities described in this report will require access to Mount Rainier National Park and surrounding Forest Service and private lands for field work, sample collection, and installation and operation of scientific instruments and telemetry equipment. Much of this land is environmentally sensitive and is designated as wilderness area. Research and monitoring activities must be designed to minimize impacts to the environment. Consultation with Park Service and Forest Service staff for work on federal land and with state personnel for work on private lands must begin at the design stage of all projects in order to assure compliance with existing regulations.

Park Service and Forest Service staff can make significant contributions to the research and monitoring efforts outlined in this report. They are in a position to notice subtle changes in the volcano that might not be apparent to visiting scientists or the general public. They can make regular visual observations of snow, ice, and rocks on the volcanic edifice; assist with the collection of data; and, where appropriate, assist with inspections and routine maintenance of instrumentation. Cooperation between researchers and Park Service and Forest Service staff is essential to the successful implementation of this project.

2

MOUNT RAINIER, ACTIVE CASCADE VOLCANO

Mount Rainier (Figure 2.1) is one of about two dozen recently active volcanoes in the Cascade Range, a volcanic arc formed by subduction of the Juan de Fuca plate beneath the North American plate. Volcanism in this arc began at least 37 million years ago and has continued intermittently to the present. During that time, numerous volcanoes have formed, flourished, died, and eroded away, generally leaving behind only those deposits in protected, low-lying areas surrounding the easily eroded cones. These deposits have been buried by younger eruptions, altered by burial metamorphism, and exposed at the Earth's surface by erosion. The modern volcanoes and volcanic fields of the Cascades, which rest on this older volcanic landscape, have formed in the past 2 million years, and mostly in the past 1 million years or less (Crandell, 1963; Crandell and Miller, 1974).

The volcanic cone, or *edifice*, has been constructed from thousands of lava flows and breccias and a few ash deposits. Some of the lava flows are more than 60 m thick at the base of the edifice, in what is now Mount Rainier National Park (Figure 2.2). Some of the breccias were deposited by moving water, but others were probably emplaced during volcanic explosions or by fragmentation of moving lava flows.

The volcano is located about 35 km southeast of the Seattle-Tacoma metropolitan area (see Figure 2.1), which has a population of approximately 2.5 million people (Figure 2.3). This metropolitan area is the high-technology industrial center of the Pacific Northwest and one of the commercial-aircraft-manufacturing centers of the United States. The rivers draining the volcano empty into Puget Sound, which has two major shipping ports, and into the Columbia River, a major shipping lane and home

FIGURE 2.1 Map of Mount Rainier and surrounding regions in the State of Washington. The approximate outline of the volcano is indicated by the solid fill in Mount Rainier National Park. The locations of Puget Sound and the Seattle-Tacoma metropolitan area are indicated by dark and light stippling, respectively. Also shown are the generalized locations of the dammed reservoirs on the White, Nisqually, and Cowlitz rivers (after Swanson and others, 1992).

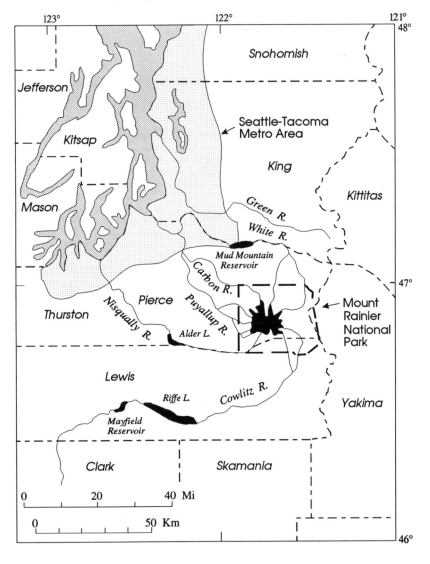

FIGURE 2.2 Generalized geologic map of Mount Rainier National Park (modified from Walsh and others, 1987).

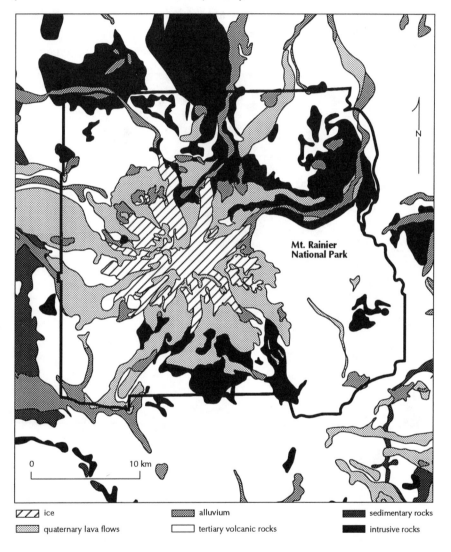

Mt. Rainier
National Park

0   10 km

| ice | alluvium | sedimentary rocks |
| quaternary lava flows | tertiary volcanic rocks | intrusive rocks |

FIGURE 2.3 Population density map of the general region shown in Figure 2.1. Incorporated areas and areas with population densities greater than or equal to 200 persons per square mile are shaded. In other areas, population densities are denoted with dots, each dot representing 100 people (courtesy of Carol Jenner, State of Washington Office of Financial Management).

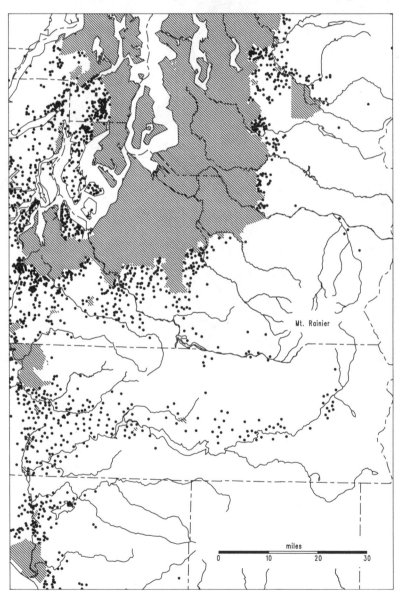

to approximately a million people in southwestern Washington and north-western Oregon.

Mount Rainier is an active volcano. It last erupted approximately 150 years ago (Mullineaux, 1974), and numerous large floods and debris flows have been generated on its slopes during this century. More than 100,000 people live on the extensive mudflow deposits that have filled the rivers and valleys draining the volcano during the past 10,000 years (Table 2.1). A major volcanic eruption or debris flow that is not prepared for could kill hundreds or thousands of residents and cripple the economy of the Pacific Northwest. Despite the potential for such danger, Mount Rainier has received little study. Most of the geologic work on Mount Rainier was done more than two decades ago. Fundamental topics such as the development, history, and stability of the volcano are poorly understood.

The recent eruptions of Mount St. Helens in the southern Washington Cascades, as well as the eruptions of other arc volcanoes such as Mount Pinatubo (Philippines), El Chichón (Mexico), Mount Unzen (Japan), and Nevado del Ruiz (Colombia), have focused the awareness of the science community, government, and the general public on volcanic hazards in the densely populated Puget Lowland area (see Figure 2.3). Public awareness of natural hazards has been heightened by the recent recognition of an active fault crossing Seattle (Atwater and Moore, 1992; Bucknam and others, 1992; Jacoby and others, 1992; Karlin and Abella, 1992; Schuster and others, 1992), which is considered capable of generating an earthquake of magnitude 7 or greater. As explained later, such an earthquake could trigger a catastrophic collapse of a portion of Mount Rainier's volcanic edifice.

Volcanic Hazards at Mount Rainier

The term *volcanic hazard* is used here to refer to a volcanic or related events that pose a threat to persons or property in surrounding regions. Table 2.2 lists 15 volcanic hazards relevant to Mount Rainier. Also shown in the table are provisional estimates of *risk*, defined as the

16

TABLE 2.1 Major Holocene Volcanic Events at Mount Rainier

Event[a]	Dates	Explosive Eruption[b]	Lahar Number	Length (km)	Reference
Glacier outburst floods and lahars	1992-1986		15	4.1	Scott and others (1992); Driedger and Walder (1991)
Lahar, Nisqually River	1970-1930		7	8	Crandell (1971)
Lahar, Tahoma Creek	1967		1	13	Crandell (1971)
Lahar, Kautz Creek	1947		2	16	Crandell (1971)
Phreatic eruption?	1894	(2-3?)[c]			Majors and McCollum (1981)
Lahar, Nisqually River	1870-1860		1	12	Crandell (1971)
Tephra X	1854-1820	(1)			Mullineaux (1974)
Lahar, West Fork White River	1695-1480		1	37	Crandell (1971); Yamaguchi (1983)
Lahar, Nisqually River	ca. 400 B.P.		1	13.5	Crandell (1971)
Lahar, North Puyallup River	> 400 B.P.		2	?	Crandell (1971)
Lahar, Tahoma Creek	ca. 440 B.P.		1	14.5	Crandell (1971)
Lahar, Tahoma Creek	470-2800 B.P.		2	10.5	Crandell (1971); Yamaguchi (1983)
Lahar, Ohanapecosh River	470-3600 B.P.		1	17.5	Crandell (1971); Yamaguchi (1983)
Lahar, South Mowich River	470-3600 B.P.		1	12	Crandell (1971)
Lahars, Kautz Creek	470-3600 B.P.		3	10	Crandell (1971); Yamaguchi (1983)
Lahar, Muddy Fork Cowlitz River	470-3600 B.P.		1	13.5	Crandell (1971); Yamaguchi (1983)
Lahar, South Mowich River	< 1480		1	11	Crandell (1971); Yamaguchi (1983)
Lahars, South Puyallup River	< 1480		4	?	Crandell (1971); Yamaguchi (1983)
Lahars, Kautz Creek	< 1480		2	11.5	Crandell (1971); Yamaguchi (1983)

Event[a]	Dates	Explosive Eruption[b]	Lahar Number	Length (km)	Reference
Electron Mudflow, Puyallup River	550-530 B.P.		1	48	Crandell (1971); Scott and others (1992)
Lahar, Nisqually River	800-3600 B.P.		7	12	Crandell (1971)
Lahar, Puyallup River	1000-1050 B.P.		1	30	Crandell (1971); Scott and others (1992)
	2200-2500 B.P.		2	10	R.P. Hoblitt (unpublished data)
Tephra C[d]	2200-2500 B.P.	1 (?)	1	24	Mullineaux (1974)
	2200-2500 B.P.		1	12	Crandell (1971)
Lahar, White River	2300-2700 B.P.		3	13	Crandell (1971)
Lahar, West Fork White River	2300-3600 B.P.		1	37	Crandell (1971)
Lahar, White River	2300-3600 B.P.		1	33	Crandell (1971)
Round Pass Mudflow, Tahoma Creek and Puyallup River	2170-2790 B.P.		1	31	Crandell (1971); Scott and others (1992)
Lahar, Nisqually River	< 3600 B.P.		1	47	Crandell (1971)
Lahar, Kautz Creek	> 3600 B.P.		1	11.5	Crandell (1971)
Lahar, South Mowich River	> 3600 B.P.		1	18.5	Crandell (1971)
Lahar, South Puyallup River	> 3600 B.P.		1	?	Crandell (1971)
Lahars, West Fork White River	3600-5700 B.P.		2	18	Crandell (1971)
Lahars, Nisqually River	3600-6800 B.P.		3	30.5	Crandell (1971); Bacon (1983)
Lahar, Carbon River	3600-6800 B.P.		1	18.5	Crandell (1971); Bacon (1983)
Lahar, Ohanapecosh River	3600-6800 B.P.		1	17.5	Crandell (1971); Bacon (1983)
Tephras B and H	3900-5000 B.P.	(2)			Mullineaux (1974)
Tephra F	4500-5000 B.P.	(1)			Mullineaux (1974)

18

TABLE 2.1 Continued

Event[a]	Dates	Explosive Eruption[b]	Lahar Number	Length (km)	Reference
Osceola Mudflow, West Fork White River	4500-5000 B.P.		1	113	Crandell (1971); Scott and others (1992)
Paradise Lahar, Nisqually River and Paradise River	4500-5000 B.P.		1	29	Crandell (1971); Scott and others (1992)
Tephras S and N	5000-5800 B.P.	(2)			Mullineaux (1974)
Lahars, White River	5700-6800 B.P.		5	13	Crandell (1971)
Tephra D	5800-6400 B.P.	(1)			Mullineaux (1974)
Tephra L	6400 B.P.	(1)			Mullineaux (1974)
Tephra A	6400-6700 B.P.	(1)			Mullineaux (1974)
Lahar, Ohanapecosh River	> 6800 B.P.		1	9.5	Crandell (1971)
Tephra R	> 8750 B.P.	(1)			Mullineaux (1974)

SOURCE: Modified from Hoblitt and others, 1987.

[a] Many smaller events are given by Scott and others (1992, Table 3 and text).
[b] Number of events. Parentheses indicate volume less than 0.1 km^3.
[c] Questionable phreatic explosions.
[d] At least one pyroclastic flow of 12 km length, and one or more lava flows of at least 3.5 km length, roughly coincide with eruption of layer C.

probability of loss of life, property, and productive capacity in the area affected by the hazard. Risk depends in part on proximity to the hazard, which is defined here in terms of distance from the volcanic edifice. The *proximal zone* refers to areas on and adjacent to the volcanic edifice. For Mount Rainier, the proximal zone generally lies within the boundaries of Mount Rainier National Park (see Figure 2.1). The *distal zone* refers to areas beyond the edifice that could be affected significantly by volcanic activity. For Mount Rainier, the distal zone includes areas up to about 100 km outside of the National Park. Risk also depends on the size (magnitude) of the event and its frequency of occurrence. In general, high-magnitude events pose greater risks to people and property than low-magnitude events. Relatively little is known about magnitudes and frequencies of volcanic events at Mount Rainier, so the estimates of risk shown in Table 2.2 are necessarily qualitative.

Studies of the geologic history of Mount Rainier and other Cascades volcanoes (see Chapter 3 in this report) suggest that major volcanic hazards are likely to include the following:

- *Volcanic eruptions.* The eruption of lava flows and tephra (particulate materials such as ash).
- *Edifice failure.* The gravitational collapse of a portion of the volcano.
- *Glacier outburst floods (jökulhlaups).* The sudden release of meltwater from glaciers and snowpack or from glacier-dammed lakes on the edifice.
- *Lahars, or debris flows, and debris avalanches.* Gravitational movement of commonly water-saturated volcanic debris down the steep slopes of the volcano and into nearby valleys.

The most likely volcanic hazards at Mount Rainier are from debris avalanches, lahars, and floods like those of the past that have repeatedly swept down the valleys heading on the edifice (Crandell and Mullineaux, 1967; Crandell, 1973; Scott and others, 1992). Frequency and magnitude estimates for such events can be made by reconstructing the spatial and

TABLE 2.2 Potential Volcanic Hazards at Mount Rainier

Hazard	Probable Risk		Need to know[a]
	Prox-imal	Distal	
Lava flows	M?	L	How far, how fast, role in producing melting of snow and ice
Phreatic and phreatomagmatic eruptions	H?	L	Generation, potential size, favored eruption site(s)
Ballistic projectiles	H	L	Size of ballistics
Tephra	H	H	Frequency of small falls
Pyroclastic flows and surges	H?	H	Frequency of occurrence; role in producing melting of snow and ice
Lahars[b]	H	H	Origin, how far, how fast
Jökulhlaups[b]	M	L	Role of heat flow in production
Sector collapse[b]	H	H	Causes, sizes
Landslides[b]	M	L	
Rock and debris ava-lanches[b]	H	L	Causes, sizes, association with steam blasts
Volcanic earthquakes[b]	L?	L	
Ground deformation[b]	L?	L	
Tsunami	L	L?	

Hazard	Probable Risk		Need to know[a]
	Prox- imal	Distal	
Airshocks	L	L	
Gases and aerosols[b]	M?	L	

NOTES: Proximal and distal refer to areas within and outside Mount
Rainier National Park, respectively.
L= low; M = moderate; H = high.
[a] With more refined information on magnitude, frequency, and areas affected for all hazards.
[b] Hazards that can occur when volcano is not in eruption.

temporal distributions of lahars and debris avalanches preserved in the stratigraphic record on and around the edifice. For example, the frequency
with which lahars have affected areas more than 20 km from the volcano
in the past (Table 2.1) suggests that the annual probability of such an event
is about 0.001. An event of this magnitude would be expected to occur an
average of once every 1,000 years. Similarly, lahars that extend to distances of 50 km or more from the volcano have an estimated annual probability
of about 0.0001. An event of this magnitude would be expected to occur
an average of once every 10,000 years. These larger lahars could affect the
Puget Lowland, inundating tens to hundreds of square kilometers in relatively densely populated areas. These probabilities should be considered as
minimum estimates, because they are based on incomplete mapping of
lahar distributions. As additional lahars are identified through field investigations, these probabilities could be revised upward. That is, these events
could be seen as occurring with greater frequency than present estimates
would suggest.

The most voluminous debris avalanches and lahars at Mount
Rainier originated from parts of the volcano that contained large volumes
of hydrothermally altered materials (Crandell, 1971; Scott and others,
1992). Frank (1985) concluded that the upper west flank and the summit

(Figure 2.4) could provide the largest sources of material for future lahars. Consequently, rivers heading on the west and northwest sides of the volcano are particularly vulnerable to large debris avalanches and lahars. These include the Puyallup and Carbon rivers, which drain into Puget Sound through the densely populated Seattle-Tacoma metropolitan area (Figure 2.1). Debris avalanches of large volume are most likely to occur during eruptions, but they could also occur during dormant periods (Crandell, 1971; Frank, 1985; Scott and others, 1992).

All the major rivers that drain Mount Rainier, except the Puyallup-Carbon system, are dammed at distances of 40 to 80 km downvalley from the summit (Figure 2.1). If reservoirs were empty or nearly so, these dams could probably contain all but the very largest expectable lahars and floods. However, if the reservoirs were full or nearly so, lahars could cause overtopping of the dams and significant downstream flooding.

Damage to property and loss of life from debris flows could be substantial. Geologic mapping of surficial deposits in Mount Rainier National Park by Crandell (1969), and later investigations by Scott and others (1992), have shown that numerous debris flows have entered the Puyallup, Nisqually, and White rivers in the past several thousand years. The largest debris flow from Mount Rainier, the Osceola Mudflow, occurred about 4,500 to 5,000 years ago (Scott and others, 1992; see Table 2.1). This mudflow traveled down the White River drainage system a distance of approximately 110 km, transported at least 3 km^3 of rock debris (Scott and others, 1992), and buried parts of the Puget Lowland that are now heavily populated (see Figures 2.5 and 2.6(A)). For comparison, the devastating debris flows of the 1985 eruption of Nevado del Ruiz that killed 25,000 people had a volume of 0.048 km^3 (Lowe and others, 1986), only about 2.4 percent of the volume of the Osceola Mudflow.

Another major lahar, the Electron Mudflow, swept down the Puyallup River Valley (see Figure 2.5) about 550 years ago. Recent excavations in this mudflow for construction in the town of Orting, about 50 km from the volcanic edifice, uncovered 2-m diameter Douglas fir stumps

FIGURE 2.4 (A) View of the west flank of Mount Rainier. The curved area of cliffs in the foreground rings Sunset Amphitheater, the source area for the Electron Mudflow. The central point at the summit is Columbia Crest, formed 2,200 to 2,500 years ago.

FIGURE 2.4 (B) View of the headwall and known site of hydrothermally altered rock below Point Success (the rightmost summit peak in (A)) showing steeply dipping lava flows, flow rubble, volcanic breccias, and tuffs extending away from the upper part of the volcano, which was removed by events that produced the Osceola Mudflow and other large lahars. (Photos courtesy of David Frank, U.S. Environmental Protection Agency.)

FIGURE 2.5 Map of Mount Rainier and surrounding regions showing the generalized locations of three large lahars indicated by dark stippling: the Osceola Mudflow (White River), Paradise Lahar (Nisqually River), and Electron Mudflow (Puyallup River). After Swanson and others (1992) with data from Crandell (1971).

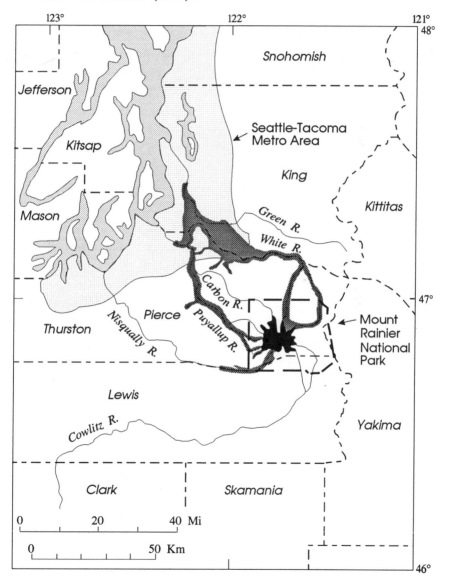

from an old-growth forest that was buried by the mudflow (Patrick Pringle, Washington State Department of Natural Resources, oral communication, 1993; see Figure 2.6(B)).

In the past 45 years, several dozen debris flows and outburst floods have occurred at Mount Rainier, the majority in the Tahoma Creek-Nisqually River drainage (Table 2.1; Scott and others, 1992). These flows did not directly threaten communities, but they did affect areas frequented by visitors to the National Park and required the expenditure of Park Service funds for cleanup and reconstruction. One popular road remains closed. The largest debris flow extended about 16 km from its origin on the volcanic edifice (Crandell, 1971). Debris flows of this magnitude are essentially unpredictable at current levels of understanding, but they serve as a reminder of the dynamic landscape surrounding Mount Rainier.

Events such as edifice failures, glacier outburst floods, and debris flows can occur in the absence of volcanic eruptions. Mount Rainier is the high- est volcano in the Cascade Range (4,392 m above sea level, with approximately 3,000 m of relief) and contains about 140 km^3 (Sherrod and Smith, 1990) of structurally weak, locally altered rock capped by about 4.4 km^3 of snow and ice (Driedger and Kennard, 1986), all of which stand near the angle of repose. The volcano is inherently unstable (Figure 2.7). Ground shaking during an earthquake could cause the gravitational failure of a large sector of the volcanic edifice, producing catastrophic avalanches and debris flows, and possibly triggering an eruption. Indeed, several large debris flows from Mount Rainier have been generated apparently without eruptive products. A pertinent example is the Round Pass Mudflow, which is approximately 2,600 years old (Scott and others, 1992; Patrick Pringle, Washington State Department of Natural Resources, oral communication, 1994).

Catastrophic edifice failure is generally recognized to be a severe hazard at stratovolcanoes such as Mount Rainier (Siebert, 1992; López and Williams, 1993). On a worldwide basis, such collapses have occurred an average of four times a century for the past 500 years (Siebert, 1992). A

FIGURE 2.6 (A) Cross-section of the Osceola Mudflow near Buckley, Washington, approximately 85 km downstream from the volcano. The mudflow, which caps the 30-m high terrace, is 3 to 5 m thick. The boundary between the mudflow and underlying glacial deposit is indicated by an arrow in the photo.

FIGURE 2.6 (B) Stump of a 2-m diameter Douglas fir uncovered during excavation near Orting, located near the confluence of the Carbon and Puyallup rivers (see Figure 2.5). The stump is from an old-growth forest that was buried by the mudflow. (Photos courtesy of Patrick Pringle, Washington State Department of Natural Resources.)

FIGURE 2.7 Interlayered lava flows and mudflows in a 30-m high cliff near the summit of the volcano. These layers probably formed when thin lava flows moved downslope, melting ice and snow to create mudflows. The rubble layers dip away from the summit at angles up to 30° and are inherently unstable (from Fiske and others, 1963, Figure 51).

particularly pertinent example of such a collapse occurred at the Bandai-san Volcano in Japan in 1888; this failure occurred without any precollapse or postcollapse eruptive activity (Sekiya and Kikuchi, 1889).

Glacier outburst floods and lahars can occur during warm summer days, heavy rainfalls, or as a result of transient heating events that melt the snow and ice cover on the volcano. Several lahars were generated on Mount Rainier in 1947 as a result of heavy rainfall (Crandell, 1971). The volcanic edifice contains a well-developed hydrothermal circulation system that transfers heat from depth to the surface (Frank, 1985, in press). This system is supplied by precipitation at the surface of the volcano, and pathways for fluid flow are provided by the numerous faults and fractures in the edifice. Changes in this "plumbing system" due to the formation of new faults and fractures could bring heated fluids into contact with snow and ice on the volcanic edifice, causing rapid melting and runoff. Such heating could occur without warning. Transient thermal events have been observed on other volcanoes in the Cascades, for example, at Mount Baker, Washington, in 1975 (Malone and Frank, 1975; Frank and others, 1977).

Eight of the Cascade Range volcanoes or volcanic fields have erupted in the past 500 years. Six of these, including Mount Rainier, have erupted in the past 200 years (Table 2.3). Four of the eruptions in the past 200 years were relatively large and could have caused considerable property damage and loss of life if they occurred today. Pyroclastic flows and lahars that entered the Columbia River were produced from Mount Hood in the 1790s and about 1800 (Cameron and Pringle, 1987). Mount St. Helens erupted explosively in late 1799 or early 1800 (Yamaguchi, 1983), producing a widespread tephra deposit (Mullineaux, 1986). In 1915, Lassen Peak in California erupted pyroclastic flows with accompanying lahars that created a "devastated" zone north and northeast of the volcano. Mount St. Helens erupted violently in 1980, killing 57 people through the combined effects of a debris avalanche formed by a giant landslide, a lateral blast expelled as the slide depressurized the volcanic system, and lahars generated by the eruption (Lipman and Mullineaux, 1981).

TABLE 2.3 Principal Volcanoes and Volcanic Fields in the Cascade
Range and Dates of Their Most Recent Volcanic Activity, Listed from
North to South

Volcano	Location	Date of most recent volcanism[a]
Silverthrone	British Columbia	Possibly younger than 1000 A.D.
Bridge River cones	British Columbia	Possibly younger than 500 A.D.
Meagher Mountain field	British Columbia	About 300 B.C.
Mount Cayley	British Columbia	About 200,000 years ago
Mount Garibaldi	British Columbia	Early Holocene
Mount Baker	Washington	1880 (1884?) A.D.
Glacier Peak	Washington	18th century(?)
Mount Rainier	Washington	1820-1854 (1894?) A.D.
Goat Rocks	Washington	Late Pleistocene(?)
Mount Adams field	Washington	Eleven Holocene eruptions, possibly none younger than 1500 B.C.
Mount St. Helens	Washington	1980s A.D.
Indian Heaven field	Washington	About 6000 B.C.
Mount Hood	Oregon	1865 A.D.
Mount Jefferson area	Oregon	About 4500 B.C.
Belknap Crater	Oregon	About 360 A.D.

TABLE 2.3
Continued

Volcano	Location	Date of most recent volcanism[a]
North Sister field (Collier Cone)	Oregon	About 950 A.D.
South Sister (south flank)	Oregon	About 100 A.D.
Mount Bachelor	Oregon	Early Holocene
Newberry Volcano	Oregon	About 620 A.D.
Mount Mazama (Crater Lake)	Oregon	About 5000 B.C.
Mount McLoughlin	Oregon	20,000-30,000 years ago
Mount Shasta	California	1786 A.D.
Medicine Lake Highland	California	About 1000 A.D.
Lassen Peak	California	1914-1917 A.D.
Cinder Cone	California	About 1650 A.D.

NOTES: See Figure 3.1 in this report for geographic reference.
[a] Sources: Simkin and others, 1981, 1984; Hildreth and Fierstein, 1983; Hoblitt and others, 1987; Harris, 1988; Souther and Yorath, 1991; Wood and Kienle, 1990; M. A. Clynne, U.S. Geological Survey, oral communication, 1992.

These examples illustrate that Cascade Range volcanoes are capable of major eruptions, especially after long periods of quiescence. About 1,000 years of quiet at Lassen Peak preceded the 1914-1917 eruption, some 200 to 400 years of inactivity predated the late-nineteenth century eruptions at Mount Hood, and some 600 years of quiescence foreshadowed the activity at Mount St. Helens that began in 1480 and continued intermittently until 1857. In fact, circumstantial evidence suggests that long periods of inactivity at some intermittently active volcanoes end with particularly violent eruptions; apparently, energy is being stored rather than released in small eruptions (Simkin and others, 1981).

Mount Rainier is capable of eruptions of moderate to very large magnitude, as measured by the Volcanic Explosivity Index (Newhall and Self, 1982) of 4 to 5 tentatively assigned to the explosive eruption that occurred between 30,000 and 100,000 years ago (Hoblitt and others, 1987). Its record of inactivity in the twentieth century and minor activity in the past few hundred years is not unusual for an active volcano. Indeed, based on past history, there is good reason to believe that the volcano will erupt again. Even relatively small eruptions could generate large floods and debris flows from melting of snow and ice on the summit. As noted previously, these debris flows and floods could cause significant property damage and loss of life along the river valleys draining the volcano, which tend to be heavily populated (Figure 2.3).

Based on the known Holocene history of the volcano, the most likely future eruptive event at Mount Rainier is the extrusion of a lava flow at the summit, possibly accompanied by tephra eruptions. Geologic mapping (Fiske and others, 1963) has documented that numerous lava flows have been erupted recently in Mount Rainier's history, and the youngest of these, stubby flows up to 60 m thick, are preserved on the floor of present-day valleys and extend only a few kilometers away from the base of the volcano. Past history suggests that future lava flows will likely be restricted to valley floors within Mount Rainier National Park or will extend only a short distance outside the park.

Although limited in areal extent, lava flows from Mount Rainier would destroy roads, buildings, and other fixed installations in and near valley bottoms and would be disruptive to many activities in the park. The sluggish motion of these flows would likely permit people to safely evacuate areas that were at risk, which means that little loss of life from this hazard would be expected.

The public perception of risks associated with lava flows, if they occurred, could be far greater than the actual risks. The glowing surface of a flow could be exposed for many days, and the nighttime reflection of this glow from the underside of weather clouds might be visible to the hundreds of thousands of residents of the Puget Lowland. Similarly, columns of steam produced by the interaction of lava flows with snow and ice on the edifice might be visible to people at great distances from the volcano. These phenomena could convey a sense of impending crisis that would not be warranted because, as noted above, lava flows are likely to be restricted to National Park land.

Explosive eruptions from Mount Rainier could send clouds of tephra high into the atmosphere, where it would be carried laterally by prevailing winds before settling to the ground. Past eruptions of the volcano have deposited up to 2.5 centimeters (cm) of ash 40 km downwind of the edifice (Crandell, 1973). Tephra would be a hazard to crops and other vegetation, machinery, and poorly built structures (the weight of the tephra could cause these structures to collapse). People living and working in such structures and those with respiratory problems would be at risk.

The prevailing winds in western Washington are from southwest to northeast, so tephra from Mount Rainier would normally be carried away from the Seattle-Tacoma metropolitan area (Figure 2.1). Less frequently, winds blow from east to west, and during these times tephra could be scattered over much of the Puget Lowland. This would disrupt commerce, travel (especially at Seattle-Tacoma International Airport), and the daily lives of hundreds of thousands of people.

Assessing frequencies and magnitudes of tephra eruptions is difficult. None of the thousands of flows on the edifice has been isotopically dated, nor is the age known for any tephra older than about 6,700 years. Ten tephras younger than about 6,700 years have been recognized and dated either directly or by bracketing between dated lahars. In postglacial times, according to Hoblitt and others (1987), the annual probability of a tephra eruption of small volume, between 0.01 and 0.1 km^3, is about 1 in 1,000. The effects of such an eruption would be minor beyond a distance of approximately 50 km from the edifice. Hoblitt and others (1987) also estimate that the annual probability of an explosive eruption producing more than 0.1 km^3 of tephra, which would have serious effects beyond approximately 50 km, is about 1 in 10,000. A task of future studies of Mount Rainier is to refine these estimates and, in particular, to estimate the average recurrence intervals for the lower magnitude but more frequent eruptions of lava flows.

Recommendations

Mount Rainier poses a significant hazard to life and property in heavily populated areas surrounding the volcano, particularly in the Seattle-Tacoma metropolitan area. The most likely hazards include edifice failures, glacier outburst floods, and lahars, with or without volcanic eruptions. Coordinated research that involves both geoscientists and social scientists should be undertaken to determine potential magnitudes and frequencies of potential hazards, their human and economic impacts, and strategies for using such information effectively to mitigate risk as part of this Decade Volcano Demonstration Project. A plan to achieve these objectives is outlined in the remainder of this report.

3

DEVELOPMENT AND HISTORY OF MOUNT RAINIER

This chapter provides a brief overview of the development of Mount Rainier from a regional perspective. It also outlines the detailed studies needed to assess potential hazards to life and property in the areas surrounding the volcano. The evaluation of hazards at Mount Rainier requires an understanding of the geologic history of the volcano and the surrounding region. Study of the geologic history is important because it is a guide, though admittedly imperfect, to what may happen in the future. A regional perspective is necessary because regional effects such as ground shaking due to large earthquakes on distant faults could have significant local consequences, as discussed in Chapter 2, and because the impacts of potential hazards such as debris flows could extend several tens of kilometers downstream of the edifice. The history of the volcano during the Holocene is particularly important, because, as explained below, this history is probably most directly relevant to assessing future hazards.

Much of the recommended work at Mount Rainier can be linked to ongoing research programs of federal, state, and academic scientists that address the broader geologic analysis of the Pacific Northwest. For example, a major investigation of crustal structure is planned in 1995-1996 by the U.S. Geological Survey (USGS) Deep Continental Studies Program along an east-west corridor extending from the Pacific coast to the Columbia Plateau. This corridor, which passes through the Mount Rainier region, will be the site of seismic refraction, wide-angle reflection, magnetotelluric, gravity, magnetic, geologic mapping, and geochemical studies. Results from these investigations can be integrated into the regional work to provide a more complete understanding of crustal architecture beneath and surrounding the volcano.

Regional Setting and History

The Cascade arc (Figure 3.1) can be divided into five segments, based on the distribution of volcanic vents formed since about 5 million years ago (Guffanti and Weaver, 1988). Mount Rainier is at the north end of a segment characterized by the relatively low production of dominantly basaltic lava, with andesite and dacite concentrated in five large Quaternary centers (Mount Hood, Mount St. Helens, Mount Adams, Goat Rocks Volcano, and Mount Rainier). According to Sherrod and Smith (1990), the highest rate of eruptive activity during the past 5 million years has been in Oregon, where an average of between 3 and 6 km^3 of lava has been erupted per million years (m.y.) per kilometer along the Cascade arc (i.e., 3–6 km^3/km/m.y.). In northern California and southern Washington, average rates of eruptive activity were 3.2 km^3/km/m.y. and 2.6 km^3/km/m.y., respectively. North of Mount Rainier, virtually all eruptive activity has been concentrated at the major composite cones of Glacier Peak, Mount Baker, Mount Garibaldi, and Meagher Mountain.

Mount Rainier occurs in a dominantly compressional tectonic setting, in contrast to the extensional setting that characterizes the Oregon and California Cascades. The current regional stress field in the crust, as determined from earthquake focal mechanisms, is roughly horizontal north-south compression (Ma, 1988; Ma and others, 1991). East-west Quaternary extension can be inferred south of Mount Rainier from the presence of locally abundant basalt, vents for which are aligned north-south in the Indian Heaven basalt field. North-striking normal faults of Quaternary age have not been identified in this area, however. East-west extension started during the middle or late Miocene in the Oregon Cascades and is apparently extending slowly northward into Washington, allowing basalt to erupt and perhaps eventually leading to formation of a graben such as that in the central Oregon High Cascades (Taylor, 1990).

FIGURE 3.1 Map showing the Cascade Arc and modern plate configuration (modified from Mooney and Weaver, 1989). Volcanoes: Si, Silverthrone; Br, Bridge River cones; Me, Meagher Mountain; Ca, Mount Cayley; Ga, Mount Garibaldi; B, Mount Baker; G, Glacier Peak; R, Mount Rainier; GR, Goat Rocks; A, Mount Adams; IH, Indian Heaven; S, Mount St. Helens; H, Mount Hood; J, Mount Jefferson; BC, Belknap Crater; TS, Three Sisters Cluster; Ba, Mount Bachelor; N, Newberry Volcano; C, Crater Lake (Mount Mazama); Mc, Mount McLoughlin; M, Medicine Lake; Sh, Mount Shasta; L, Lassen Peak; CC, Cinder Cone.

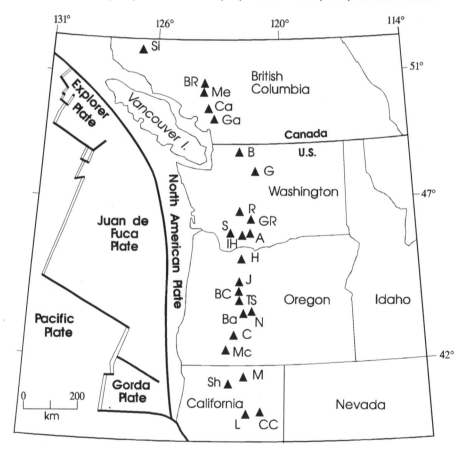

The large volcanic centers of Mount Rainier, Mount Adams, Goat Rocks, and Mount St. Helens (Figure 3.2) form a triangular arrangement that is unique in the Cascades. These centers lie along the edges of a major midcrustal electrical conductivity and magnetic anomaly that may represent sedimentary rocks deposited in a marine forearc basin and thrust against a Cretaceous-age (approximately 65-million to 140-million-year-old) continental margin, according to Stanley and others (1987, 1992). This anomaly is labeled "SWCC" in Figure 3.2. Mount Rainier, Mount Adams, and Goat Rocks volcanoes are probably located along the Late Cretaceous continental margin that forms the east side of the SWCC; Mount St. Helens occurs on the western margin of the SWCC.

Major fault systems in the region (see Figure 3.2) show no evidence of current seismicity. Twenty years of monitoring by the Washington Regional Seismograph Network show that seismicity borders the southwest side of the SWCC along the St. Helens seismic zone (SHZ; Weaver and Smith, 1983), the northwest side of the SWCC along a north-south zone in the western Rainier seismic zone (WRSZ) about 15 km west of Mount Rainier, and the east side of the SWCC along a poorly defined zone of earthquakes near Goat Rocks Volcano (Figure 3.2). A cluster of earthquakes is also located directly under Mount Rainier; this cluster is so shallow that it is probably related to small stress changes in the magma-conduit system rather than to regional stress (Malone and others, 1991).

Earthquakes up to magnitude 4 are frequent in the WRSZ, and a 5.5-magnitude event occurred just south of this zone in 1987, near Storm King Mountain between Morton and Elbe, about 35 km from Mount Rainier. The length of the WRSZ (25 km) is such that an earthquake of magnitude 6 or larger is possible if a single fault plane extends that distance (Weaver and Smith, 1983); large stresses would be placed upon the edifice by ground shaking during such an earthquake, and this could lead to slope failure or large debris flows along major drainages, as discussed previously.

Geologic observations suggest considerable Neogene uplift of the Mount Rainier area. Structure contour maps suggest that the Columbia River Basalt Group (CRBG) was uplifted more than 1.5 km along the Cascades at the general latitude of Mount Rainier and Goat Rocks volca-

FIGURE 3.2 Map of Mount Rainier and surrounding areas showing the locations of major crustal features: WRSZ, western Rainier seismic zone; SHZ, St. Helens seismic zone; SWCC, southern Washington Cascades conductor; SCF, Straight Creek Fault (inactive); N, Naches fault zone (also inactive); the Seattle Fault was recently discovered and is thought to be active. The dashed box shows the approximate area for the recommended regional studies.

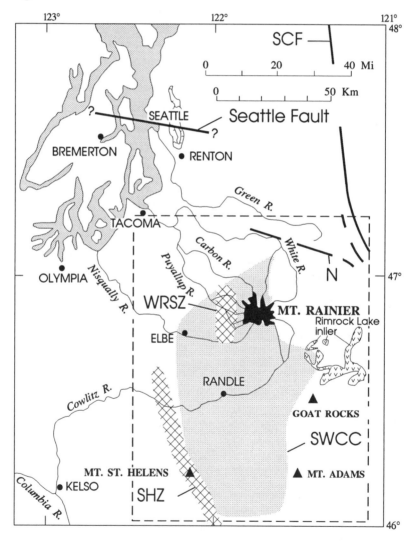

noes (Swanson and others, 1979; unpublished field data of D. A. Swanson). The amount of uplift decreases south of this latitude and probably northward as well, although the entire Cascade Range in Washington experienced some uplift. For example, Tolan and Beeson (1984) have suggested that the crest of the Cascades on the Oregon-Washington border may have been uplifted 500 to 600 m in the past 2 million years. Uplift of the Mount Rainier area took place in the past 15 million years, probably mostly in the past 12 million, based on the age of uplifted units of the CRBG. The presence of the coarse-grained mid-Miocene Tatoosh pluton directly beneath Mount Rainier suggests uplift and erosional unroofing of 1 km or more.

Uplift of the volcano may be due to magma injection in the middle and upper crust, although some uplift may be related to subduction processes that raised the Olympic Mountains beginning about 12 million years ago (Brandon and Calderwood, 1990). In order to interpret local evidence for uplift, it is essential to investigate other uplift and deformation components such as those related to retreat of Puget Sound glaciers and tectonic deformation of the arc region caused by subduction. Such investigations inherently involve a broader look at the southwestern Washington region.

Development of the Volcanic Edifice

The geology of Mount Rainier's edifice has received little study since the work of Fiske and others (1963). That research defined the geologic framework of Mount Rainier National Park, and to some extent the stratigraphy north and east of the park (Waters, 1961), but was not sufficiently detailed to reveal (1) the eruptive history of Mount Rainier proper; (2) structural features of the volcanic edifice, including small-scale faulting and dike swarms; and (3) the distribution of hydrothermally altered, structurally weakened rocks. These aspects of the geology are particularly important when evaluating volcanic hazards.

Development of the Mount Rainier Volcano probably began in the early or middle Pleistocene (i.e., 0.7 million to 1.7 million years ago;

Crandell, 1963; Crandell and Miller, 1974). Multiple lahars and layers of tephra from Mount Rainier are interbedded with glacial deposits in the adjacent Puget Lowland. Clasts in the lahars, as well as the tephra, contain hornblende phenocrysts and are unlike most products of the modern volcano. These volcanic rocks, assigned to the Lily Creek and Puyallup Formations, have reversed magnetic polarities and are older than an 840,000-year-old layer of volcanic ash (Easterbrook and others, 1981, 1985). These rocks record the early eruptions of Mount Rainier or its ancestor.

The base of the modern volcano overlies a rugged surface eroded into Tertiary-age (1.7-million to 65-million-year-old) rocks of the Cascade volcanic arc. Most of Mount Rainier's cone was built by lava flows of intermediate composition interbedded with breccia and minor tephra, including pumice. High on the cone, the flows are rarely more than 15 m thick. Many flows thicken downslope and are more than 60 m thick on the lower slopes of the cone, where they partially fill paleovalleys (Fiske and others, 1963). The paleovalleys are eroded into the basement rocks radial to the volcano; an earlier cone may have conditioned the development of this radial drainage. Much of the breccia on the cone was probably produced by the interaction of lava flows with snow or glacier ice or from autobrecciation of the flows themselves, but some of the breccia was probably derived from explosions and lahars. Radial dikes, which may have fed some of the flows, are prominent in places (Smith, 1897; Coombs, 1936; Fiske and others, 1963). Little petrographic or chemical study has been made of Mount Rainier's rocks; what has been published (Coombs, 1936; Fiske and others, 1963; Condie and Swenson, 1973) indicates petrographically uniform two-pyroxene andesite and basaltic andesite. Dacite and basalt are present but apparently uncommon.

The flows and breccias eventually built a cone standing between 2,100 and 2,400 m above its surroundings before the end of the last major glaciation about 10,000 years ago. Two late Pleistocene vents erupted more mafic olivine-phyric basaltic andesite near the northwest base of the cone after Mount Rainier was almost fully developed (Fiske and others, 1963). Additional lava flows may have been added to the volcano in the late Pleistocene and early Holocene, but they have not been dated.

The largest known explosive eruption of Mount Rainier occurred between about 30,000 and 100,000 years ago and is recorded by a pumice deposit that has been recognized northeast, east, and southeast of the volcano. The deposit is about 2 m thick at a site 12 km northeast of the present summit (D. R. Crandell, written communication, cited in Hoblitt and others, 1987). Its distribution and thickness farther east are not known, nor is it known whether the thickness of 2 m is uniform along the axis of the deposit. This observed thickness at a distance of 12 km is greater than that of tephra layer Yn at a similar distance from Mount St. Helens (Mullineaux, 1986) but less than that of layers B and G from Glacier Peak (Porter, 1978). Layers Yn, B, and G all have estimated volumes equal to, or greater than, 1 km^3 (Crandell and Mullineaux, 1978; Porter, 1978). The limited thickness data for the tephra layer at Mount Rainier suggest that it may have a comparable volume. The volume of this late Pleistocene tephra is probably at least an order of magnitude greater than that of the most voluminous tephra layer of postglacial age.

The andesitic summit cone (Columbia Crest; Fiske and others, 1963, Figures 49 and 55; see also Figure 2.4 (A) of this report) is a late Holocene feature, built during the same period as the eruption of tephra layer C between 2,200 and 2,500 years ago. This cone may have developed in a crater produced by the edifice collapse that generated the Osceola Mudflow (Crandell, 1969; Mullineaux, 1974). This well-preserved cone stands about 250 m above the present crater rim. Two small craters indent the top of the cone; their rims are commonly free of snow as a result of fumarolic activity.

Holocene eruptive activity at Mount Rainier (Table 2.1) produced 11 tephra layers ranging in estimated volumes from 0.001 to 0.3 km^3 (Crandell and Mullineaux, 1967; Crandell, 1969; Mullineaux, 1974). The most voluminous tephra, layer C, has blocks and bombs as large as 30 cm in diameter at a distance of 8 km from the summit of the volcano. This layer is about 15 cm thick at a distance of 12 km east of the summit, and 8 cm thick at a distance of 25 km from the summit. Mullineaux (1974) estimated its minimum uncompacted volume to be about 0.3 km^3. A possibly correlative hot block-and-ash flow was emplaced west of the volcano in the South Puyallup Valley (Crandell, 1971). In addition to the Mount

Rainier tephra, ash from Mount Mazama (Crater Lake) and Mount St. Helens is widespread within the National Park (Mullineaux, 1974).

The eruptive products of Mount Rainier have been characterized petrographically as chemically homogeneous two-pyroxene andesite that commonly contains traces of olivine and very rarely hornblende (Fiske and others, 1963, 1964). However, few samples were examined and no new geochemical data were presented by Fiske and his colleagues. To date, only 15 chemical analyses, ranging from medium-K silicic andesite to low-silica dacite, have been published; 5 of those analyses date from the 1930s (Coombs, 1936). In a more recent study of tephra deposits from Mount Rainier, Mullineaux (1974) showed that magmas encompassing a larger diversity in phenocryst populations, whole-rock compositions, and melt compositions, including probable basaltic liquids, have been erupted from the volcano. Thus, the true abundances of magma types erupted from Mount Rainier and their associated eruptive behavior are poorly known. Two-pyroxene andesite clearly predominates, but more primitive as well as more evolved lavas are present.

During the Holocene, more than 60 debris avalanches and lahars swept down valleys heading on the volcano; the largest, the Osceola Mudflow (Table 2.1), reached an arm of Puget Sound more than 100 km from the summit (Figure 2.5; Crandell, 1971; Scott and others, 1992). Some of the debris flows contain relatively abundant clay-sized material, which evidently was derived from hydrothermally altered rocks excavated by phreatic or phreatomagmatic explosions or by the gravitational collapse of a sector of the edifice weakened by hydrothermal alteration (Frank, in press). Deposits of several lahars suggest that large collapses (>0.2 km^3) of hydrothermally altered rock debris have occurred at least three times in the past 5,000 years. Recent work by Scott and others (1992) has verified the presence of clay-rich debris flows and describes the important rheologic distinctions between clay-rich ("cohesive") and clay-poor ("noncohesive") debris flows. Scott and others (1992) ascribe the clay-rich debris flows to edifice failure of hydrothermally altered rocks.

Mount Rainier has an active hydrothermal system that perpetuates snow-free areas near the summit (Figure 3.3) and leaks fluid laterally.

FIGURE 3.3 Snow- and ice-free area containing fumaroles and heated ground on the rim of West Crater near the summit of Mount Rainier. A thermometer, shown for scale, is in a fumarole. Hydrothermal activity in this area has formed clay-rich deposits across an area spanning 1 km. (Photo courtesy of David Frank, U.S. Environmental Protection Agency.)

Frank (in press) used the chemistry of thermal fluids to infer an acid sulfate-chloride-type water as the parent water in the hydrothermal system. Weak summit fumaroles are at 86° C (the boiling-point for this fluid at the summit elevation), and Frank (in press) found that gas from the fumaroles is largely air. The significant component of hydrothermal clay in the cohesive lahars indicates that a hydrothermal system has been active within the edifice for at least 5,000 years.

Regional Studies to Assess Volcanic Hazards

The *Mount Rainier region* as defined for the purposes of this report extends from the Seattle-Tacoma metropolitan area on the north to the Columbia River on the south, and includes Mount Rainier, Mount St. Helens, and Mount Adams (Figure 3.2). A coordinated program of geologic and geophysical research within this region would address several important questions related to potential volcanic and associated hazards from Mount Rainier. Questions that need to be addressed include:

- What are the tectonic processes that control locations of volcanoes in this region? In particular, what are the relationships between volcanoes, subduction, and major crustal accretionary boundaries?

- What is the stress field in this region and how does it affect volcanism and seismicity near Mount Rainier? In particular, are there strike-slip zones that intersect the volcano, and, if so, what is their relationship to strike-slip activity along the SHZ and the WRSZ?

- What is the deformation field in the region caused by magma injection, oblique subduction, and glacial unloading?

- What are the ages, distributions, and characteristics (volumes, thicknesses, emplacement velocities) of lavas, tephras, and lahars from the volcano?

These questions can be addressed through a program of regional geologic mapping, petrologic and geochemical studies, geophysical studies, and drilling, as discussed in the following sections.

Geologic Mapping

At present, bedrock geologic information in the Mount Rainier region is inadequate to understand the history of the volcano. The reconnaissance geologic map of Mount Rainier National Park by Fiske and others (1963) has not been superseded by more detailed geologic mapping. Recent geologic map compilations of the southwest quadrant of the Washington State geologic map (Walsh and others, 1987; see Figure 2.2) provide an improved regional framework but lack detail. A strip of 7.5-minute quadrangles in an east-west zone south of Mount Rainier is currently being mapped by geologists from the USGS (for example, Evarts and others, 1987; Swanson and Evarts, 1992). This mapping complements the proposed USGS Deep Continental Studies Program referred to earlier in this chapter. In addition to this transect mapping, geologic mapping by faculty and students at Portland State University is proceeding east of Mount Rainier in the Fifes Peaks and Bumping River area (Brunstad and others, 1992; Hammond and Cole, 1992).

Detailed geologic mapping of Mount Rainier is necessary to reconstruct the history of the volcano, determine its structure and stability, and elucidate eruption and edifice failure processes. Features of particular interest include extrusive and intrusive rocks, faults, hydrothermal alteration zones, springs, fumaroles, glaciers, and surficial deposits. Geologic mapping of the volcano must be supported by isotopic dating to test geologic correlations and establish the chronology and rates of edifice growth and destruction. Paleomagnetic studies will also be required to evaluate geologic correlations, define eruptive sequences, and constrain emplacement temperatures of fragmental deposits.

Petrology and Geochemistry

Petrologic studies are needed to reconstruct the evolution of the magma system through time and to evaluate the physical characteristics of the magma and its transport from depth to the surface. These studies are important for identifying petrologic or petrochemical cycles that may culminate in explosive volcanism and for providing basic knowledge on the development of a large composite cone. The results from such studies would apply to other arc volcanoes, both in the Cascades and elsewhere. The petrologic knowledge of Mount Rainier clearly lags far behind that of neighboring volcanoes, such as Mount St. Helens (Smith and Leeman, 1987, 1993) and Mount Adams (Wes Hildreth, U.S. Geological Survey, unpublished data).

Sampling for petrologic studies must be coordinated with geologic mapping so that samples represent explicitly mapped geologic units. Petrochemical studies can be used to help establish the stratigraphic relations among flows making up the edifice. This type of systematic approach has recently been applied successfully to Mount Jefferson in the Oregon Cascades (Conrey, 1991). Samples selected on the basis of geologic context and the petrographic and compositional reconnaissance can be subjected to more detailed geochemical, isotopic, and phase-chemistry studies; techniques of experimental petrology can be applied to evaluate magma and magma-reservoir properties, including such parameters as reservoir depth and magma temperature, viscosity, and rise rate. Petrographic and whole-rock geochemical reconnaissance (such as x-ray fluorescence for major and trace element analyses) need to be carried out on all such samples.

Chemical differences among eruptive units are typically very subtle, and experience shows that geochemical methods are most useful when two conditions are satisfied: (1) that many analyses for major and trace elements are available, and (2) that all analyses are performed under identical laboratory conditions. Several different investigators and laboratories will probably be involved in studies of Mount Rainier, so interlaboratory standards and comparisons must be developed and maintained to ensure maximum utility and reliability of analytical data.

Rather extensive geochemical data on Quaternary volcanic rocks in the region exist (for example, Clayton, 1983; Hammond and Korosec, 1983; Korosec, 1989; Leeman and others, 1990; Wes Hildreth, unpublished data for Mount Adams), including some trace-element data and isotopic data, chiefly Sr-isotope ratios. Quaternary basalt is of two types, tholeiitic and calc-alkalic, but basaltic andesite to high-SiO_2 dacite are generally calc-alkalic. A few basalt flows are alkalic, some highly so.

The geochemistry of Tertiary volcanic rocks is less well characterized, but major-element data for mapped quadrangles (for example, Evarts and others, 1987; Swanson, 1992) indicate that Tertiary units are a mixed tholeiitic and calc-alkalic assemblage independent of SiO_2 content and range from basalt to rhyolite. Better regional knowledge of the Tertiary rocks is needed, because of the insights that magma compositions, geobarometry, and mineral chemistry can provide to address crustal composition, thickness, magma storage, and magma genesis. For example, major-element analyses in areas currently being mapped suggest that the crust thickened substantially during growth of the Cascade arc (Swanson, 1992). This hypothesis needs to be tested. The improved coverage must be tightly controlled stratigraphically and must include trace element and isotopic information.

Few isotopic and fission-track ages are available for pre-Holocene rocks and unconsolidated deposits throughout the entire study area (Mattinson, 1977; Hammond and Korosec, 1983; Evarts and others, 1987; Vance and others, 1987; Korosec, 1989). The overall scarcity of ages reflects both the difficulty in obtaining reliable ages for the altered Tertiary rocks and the lack of concerted effort to do so. Radiometric ages are essential for understanding recurrence intervals of volcanic eruptions at Mount Rainier and in the surrounding region. For Tertiary rocks, paleomagnetic techniques may be useful for placing constraints on ages. Improved age control for rocks in the region is an important goal of this Decade Volcano Demonstration Project.

Geophysics

Geophysical studies, such as seismic surveying, gravity and aeromagnetic mapping, electrical surveying, and ice-penetrating radar surveying, will provide data to define the following elements of the volcano:

- distributions of hydrothermally altered rock, intrusive bodies, and faults;
- presence and distribution of magmas;
- distribution of hydrothermal and groundwater systems; and
- distribution of glacier ice and the configuration of the subglacier surface.

In addition, measurements of precipitation and stream and spring discharges combined with chemical and isotopic studies will provide important constraints on the hydrologic budget and storage and transport of water within the volcano.

Seismic studies are essential to assess potential earthquake-generated hazards, including eruptions, edifice failure, and glacier outburst floods. The influence of seismic activity on edifice stability must be evaluated carefully because of the sporadic occurrence of large crustal earthquakes in southern Puget Sound, such as the 7.1-magnitude (1949) event that occurred at a depth of 25 km and the 6.5-magnitude (1965) event that occurred at a depth of 60 km. Recently discovered evidence of widespread, abrupt coastal subsidence in Washington and Oregon, and comparison of the present Cascadia subduction zone with other Pacific rim zones, together suggest that large subduction-zone earthquakes must be considered in hazard assessments for the Mount Rainier region (Atwater, 1987; Heaton and Hartzell, 1987). Atwater (1987) and Heaton and Hartzell (1987) suggest that the magnitude of such an earthquake could exceed 8.5, compared to the 7.1 subduction-zone event that occurred in 1949 as noted above, which is the maximum recorded earthquake in the region.

Regional seismic studies, as well as seismic observations on the edifice itself, utilize the University of Washington's regional seismic network, which was expanded considerably after the Mount St. Helens eruption. This expanded network has allowed detailed investigations of the WRSZ and other features that relate seismicity and tectonics in the region. In addition, data from the network have been used to show that many of the seismic events occurring on the edifice are due to glacier movement, outburst floods, and rain-triggered debris flows. To date, no low-frequency earthquakes characteristic of magmatic activity have been recorded at Mount Rainier. The network has also been used to perform tomographic imaging using teleseisms; these images show a crustal low-velocity zone that is spatially coincident with features mapped from magnetotelluric surveys. The tomographic studies also show a poorly resolved, low-velocity zone beneath Mount Rainier at depths of 9 to 25 km that may represent a magma body.

Additional tomographic studies of this kind are needed but will require increased seismometer density on the volcano to improve resolution of low-velocity "magma" zones. Demonstration of the usefulness of tomographic studies was provided at Mount St. Helens, where a high degree of detail in the subsurface was obtained with tomographic imaging using local earthquakes recorded on the local seismic network (Lees, 1992). Additions to the network are also needed to further investigate the WRSZ. This network must be expanded to cover an area extending 40 to 50 km from the volcano so that a 20-km target width can be analyzed.

An accurate seismic velocity model of the volcano and of the crustal rocks below is essential to locate seismic events of all types accurately. The existing velocity model is based on studies remote from the edifice and upon inferences about the velocities and structure of the volcanic rocks of the edifice itself. Additional data from both active and passive experiments on and around the volcano are needed.

Regional neotectonic and geodetic studies in the area shown in Figure 3.2 are largely nonexistent. Local horizontal-control networks exist at Mount St. Helens, along the SHZ north of Mount St. Helens, and at Mount Rainier. Expansion of regional surveys in western Washington using GPS (Global Positioning System) is planned by USGS workers in 1994. This expansion will extend the GPS stations into southwestern Washington, including the Mount St. Helens region. This network needs to be expanded to the north and east to encompass the regional framework of Mount Rainier as well as the volcano itself.

Regional deformation and uplift due to magmatic intrusion or tectonics needs to be addressed, possibly with the aid of regional geomorphic studies and GPS measurements. Geodetic baselines need to be established across suspected strike-slip fault zones and over possible neotectonic features such as the northwest-trending anticline located just west of Mount Rainier. Geologic mapping of brittle-fracture indicators in surficial volcanic rocks and stress studies in boreholes must be a part of neotectonic research in the region, especially west and southwest of Mount Rainier in the area of current seismicity.

Heat flow in the region has been mapped in moderate detail (Blackwell and others, 1990). A heat-flow high occurs along the western flank of the Quaternary Cascades in Oregon and extends into the southeastern quadrant of the region outlined in Figure 3.2. A heat-flow profile has been completed along an east-west corridor south of Mount Rainier, but more measurements are required north of the Cowlitz River in the study area of Figure 3.2. Temperature and heat-flow measurements within the National Park would be useful for studying hydrothermal activity on and around the edifice.

Regional seismic-velocity information includes data from a north-south refraction/wide-angle reflection profile between Randle, Washington and the Canadian border (Luetgert and others, 1992). The crustal thickness determined from this survey is approximately 44 km (Walter Mooney, USGS, oral communication, 1992) and is approximately constant beneath the Cascades (Mooney and Weaver, 1989). Extensive deep reflection surveys, sponsored by the U.S. Department of Energy, using vibrating sources have been completed on five profiles within the region; however, these data do not allow mapping of the crust below about 8 km because of poor data quality at longer recording times (>3-4 s). In addition to the deep seismic profiling planned as part of the USGS Deep Continental

Studies Program, similar refraction/wide-angle surveys may be required across the WRSZ and along profiles north of Mount Rainier. The large shots required for any of the future deep seismic surveys must be deployed so that they can be used both for calibration of the network and tomography of the volcano, utilizing the expanded seismic network and portable instruments deployed at the time of the surveys. High-resolution seismic methods are needed to study shallow fault zones and neotectonic features.

Gravity, magnetic, and magnetotelluric (MT) or deep electrical-sounding surveys have been valuable for studying the regional setting, but more work is needed around the volcano to place them in a regional context. For instance, the strong, northwest-trending gravity gradient near the center of the volcano (Finn and others, 1991) may be an expression of the interpreted major crustal boundary mapped with MT surveys. However, gravity control is poor within the National Park. More detailed coverage is needed on the Mount Rainier edifice, as well as in surrounding regions, particularly in the WRSZ. Aeromagnetic data are adequate for both Mount Rainier and its surroundings. Gravity and magnetic data have been used at other Cascade volcanoes to locate buried intrusive bodies. Examples include Goat Rocks Volcano (Williams and Finn, 1987) and Mount St. Helens (Finn and Williams, 1987; Williams and others, 1987).

MT survey coverage is relatively good on a broad regional scale (Stanley and others, 1992), but more data are needed in the region southeast of Mount St. Helens, north of Mount Adams, in the area of the WRSZ, and within the National Park. A recently acquired detailed MT profile across the WRSZ that extends into the western part of the National Park indicates that the regional deep crustal conductor (SWCC) is highly constricted to the width of the Carbon River anticline that encompasses the WRSZ (W. D. Stanley, USGS, oral communication, 1993). Additionally, an earthquake swarm in July and August 1988 appears to be related spatially to an intrusive body (probably Miocene in age) within the Eocene marine shale that has been interpreted to constitute most of the regional electrical conductor (Stanley and others, 1992). The conductor was traced to within 10 km of Mount Rainier at the east end of the MT profile.

Detailed MT and other electromagnetic surveys within the National Park would help locate alteration zones (electromagnetic methods are very sensitive to hydrothermal alteration) that represent potential sites of edifice failure, as well as zones of faulting and other hydrothermal systems on a regional scale. Several northwest-trending alteration zones occur north of the National Park. Electromagnetic surveys could help resolve whether these zones are expressions of strike-slip faults that extend southward beneath the volcano.

Arrays of portable, three-component magnetometers can be used for electromagnetic imaging of the deep electrical structure of the volcano, including magma bodies. This technique, known as geomagnetic depth sounding, or magnetovariation sounding, has the advantage of not requiring measurements of the electrical field. Such measurements would be difficult on many parts of the volcano because of the long wires needed for connecting electrodes to a data logger. Magnetometer-array studies were the first to locate the SWCC (Law and others, 1980) and could be used effectively in a combined regional and detailed study of Mount Rainier and its tectonic setting. A broad variety of electromagnetic methods, coupled with geodetic and neotectonic research, provides a good regional method of locating such faults.

Remote-sensing techniques, including multiband spectral imaging from satellite and aircraft platforms, can be used to study the edifice and surrounding region. Infrared imaging of the volcano has been used to map thermal areas at the summit, along fractures, and at other rock outcrops, as first shown by Moxham and others (1965). Multiband, near-infrared, and other remote-sensing data would be useful for mapping alteration patterns on the edifice, some of which might be related to strike-slip faulting and other neotectonic features in the region. In particular, multichannel spectral data obtainable with new aircraft-based instrumentation have the potential for discriminating alteration minerals in great detail. Such data could provide even more specific mineralogic information on surficial rocks and zones of alteration. Side-looking radar (SLAR) data may be useful for locating and mapping lineaments and faults.

Drilling

The evaluation of data and testing of models developed from the regional investigations discussed previously may require drilling. For neotectonic and fault studies, drillholes would be relatively shallow and could be used for multiple purposes, such as heat-flow measurements, borehole-seismometer installation and in situ stress measurements. A plan for deep drilling (5 to 10 km) in the western part of the SWCC and on the SHZ was presented several years ago to DOSECC (Deep Observation and Sampling of the Earth's Continental Crust, a government-funded university consortium for scientific drilling), but the project was not approved. A carefully designed science plan to drill a specific target could greatly advance understanding of the magmatic and tectonic processes at Mount Rainier and would be a logical follow on to the studies outlined in this report.

Hazard Studies

In addition to the general studies outlined above to elucidate development of the edifice, special attention needs to be focused on understanding lahars and edifice collapse, two hazards of particular significance for residents in surrounding areas.

Lahars

Lahars probably constitute the greatest hazard to life and property in the Mount Rainier region, and consequently they are an important focus of research. Recent work led to the discovery of several previously unknown lahar deposits (Scott and others, 1992). This work has demonstrated the need for detailed mapping of lahar deposits to establish a complete history of events. Correlations using cuttings from boreholes in adjacent lowland areas would help identify the downstream extent of lahars and lahar runouts. Such information would also improve volumetric estimates of alluviation during single lahar events and eruptive episodes and would allow better evaluation of liquefaction susceptibility of such deposits (Palmer and others, 1991; Pringle and Palmer, 1992).

Wood, which is commonly preserved in lahar deposits, can be used for dating and correlation studies through high-precision radiocarbon analyses and dendrochronology. For events of the past few centuries, it may be possible to achieve calendar-year accuracy using dendrochronologic methods, as Yamaguchi (1983, 1985) demonstrated at Mount St. Helens. A chronology developed using a combination of these techniques would be useful in discriminating separate events in the geologic record and in correlating among different types of deposits, such as pyroclastic flows and lahars produced during the same eruptive episode. An improved chronology might also allow mass movements on the volcano to be correlated with paleoseismic events in the Pacific Northwest (for example, Atwater, 1987; Bucknam and others, 1992; Karlin and Abella, 1992).

Edifice Collapse

Most of the geologically or hydrologically hazardous events—eruptions, collapses, and slope failures—have originated, and are likely to originate in the future, from the upper parts of the edifice. The large volume of ice and snow can amplify the destructive potential of these events. The evaluation of hazards that originate from Mount Rainier requires a comprehensive investigation of the edifice, including the following topics:

- history of growth and failure of the edifice;
- present structure and stability of the edifice and ice cap;
- the evolution and present state of the magma-supply system;

- the distribution and transport of water in the edifice; and
- the roles of water, ice, and hydrothermal alteration in past eruptions and gravitational failures.

Some portions of the volcanic edifice are more likely than others to fail by gravitational sliding or collapse along preexisting structural weaknesses such as faults or dike swarms. Dikes have been mapped on Puyallup Cleaver, along the arête on the east edge of Winthrop Glacier, and on Little Tahoma Peak. Fiske and others (1963) note that dikes are more abundant than is indicated on their geologic map. The distribution and orientation of faults and dikes, important for assessing the structural integrity of different parts of the volcanic edifice (Fink, 1991), can be determined only by more detailed geologic mapping.

A distinctive characteristic of the huge Osceola Mudflow that is indicative of its origin is its relatively high content of clay-size material, thought to be mostly clay minerals. Mullineaux (1974) found that an ashfall deposit contemporaneous with the Osceola Mudflow, the F tephra, contains both fresh, newly erupted magma and abundant clay minerals. This finding suggests that much of the clay in the Osceola Mudflow originated from hydrothermally altered rocks derived from the edifice by gravitational collapse or volcanic explosions during an eruption of juvenile magma (such as the eruption of Mount St. Helens in May 1980; for example, Janda and others, 1981; Voight and others, 1981). Such altered rocks are weaker than fresh lava flows and are thus more prone to failure. The distribution of hydrothermally altered rocks on Mount Rainier is clearly significant for assessing the structural integrity of the edifice and can only be determined by detailed geologic mapping and sampling. As previously noted, remote sensing can aid the mapping of alteration zones. The physical properties of hydrothermally altered rocks can be assessed through quantitative mineralogical and mechanical analyses.

Hot springs and fumaroles are surface manifestations of the hydrothermal system within the edifice, which is the primary agent of rock alteration and a major influence on slope stability. The process and extent of wallrock alteration by hot gas and water require better delineation, particularly in the context of the extensive glacier mantle on the volcano. Glaciers may hide significant areas of alteration, and their presence may greatly enhance rock alteration by trapping hot gas and allowing the substrate to "stew in its own juice" (Carrasco-Núñez and others, 1993). Understanding the fundamentals and areal extent of this alteration process constitutes one of the most important tasks of Decade Volcano research at Mount Rainier.

Recommendations

Coordinated research of Mount Rainier and the surrounding region, shown in Figure 3.2, should be undertaken as part of this Decade Volcano Demonstration Project. Regional studies are needed to address the formation and subsequent development of Mount Rainier within the Cascade volcanic arc environment. Of particular importance in this context are studies of the following:

• tectonic processes that control the locations of volcanic vents;
• regional stress fields and their effects on volcanism, faulting, and seismicity;
• the crustal deformation field caused by magma injection, subduction, and glacial loading; and
• ages, distributions, and characteristics of tephras, lavas, and lahars.

Studies of Mount Rainier itself are also needed to address the development of the edifice in order to predict its future behavior. Of particular importance are studies that address these topics:

• the structure of the volcanic edifice and underlying crust, including distributions of magmas, intrusive and extrusive rocks, faults, dikes, and glacial ice;

- the history of edifice growth and failure;
- the geometry of hydrothermal and groundwater systems; and
- distributions of hydrothermally altered rocks.

A high degree of feedback between local and regional studies and between individual projects such as geologic mapping and geophysical surveys should be employed as part of the strategy for this Decade Volcano Demonstration Project. This project should be coordinated with ongoing research programs of federal, state, and academic scientists and should include the following elements:

1. Geologic mapping. Mapping the spatial and temporal distributions of eruptive and intrusive rocks, faults, hydrothermal alteration zones, surficial deposits, springs, fumaroles, and glaciers should be undertaken as part of the effort to understand the development of the Cascade volcanic arc and Mount Rainier edifice. This mapping work should be supported by dating and paleomagnetic studies to establish correlations, chronologies, and rates of edifice growth and failure. Geologic mapping is particularly needed in and adjacent to the WRSZ and Mount Rainier National Park.

2. Petrologic and geochemical studies. Petrologic and geochemical (including isotopic) studies of Tertiary and Quaternary (particularly Holocene) rocks should be undertaken to address the physical characteristics and evolution of the magma systems through time, to help establish stratigraphic relations among eruptive products, and to provide the basis for reconstructing patterns of hydrothermal alteration. The isotopic work should include radiometric dating to establish recurrence intervals for volcanic eruptions at Mount Rainier and adjacent volcanoes. This petrologic and geochemical work should be coordinated with mapping efforts so that samples can be tied to explicitly mapped geologic units. In support of this effort, interlaboratory standards should be developed and maintained to establish the high level of analytical control needed to detect the subtle chemical variations expected between eruptive units.

3. Geophysical surveys. Geophysical surveys should be undertaken to elucidate the structure of the volcanic edifice and underlying crust, including distributions of magmas, intrusive bodies, faults, hydrothermal and groundwater systems, and glacier ice. Several types of surveys should be carried out to support this effort:

- *Earthquake surveys,* utilizing the regional seismic network to investigate patterns of seismicity and to obtain tomographic images of the crust beneath the edifice. Selective additions to the regional seismic network should be made to improve earthquake detection in poorly covered areas east and south of Mount Rainier and to improve the spatial resolution of deep-seated velocity contrasts for tomographic studies. A tomographic investigation of the type recently completed at Mount St. Helens (Lees, 1992) should be undertaken for Mount Rainier.

- *Seismic surveys,* particularly refraction/wide-angle reflection surveys across the WRSZ and in the region north of Mount Rainier, to elucidate crustal structure; high-resolution seismic surveys, to study shallow fault zones and neotectonic features.

- *Geodetic surveys,* utilizing GPS to monitor deformation of the region. As part of this effort, the existing geodetic network should be expanded to cover the entire region outlined in Figure 3.2 and should include an array of stations on the edifice.

- *Heat-flow surveys* in the Cowlitz River area (Figure 3.2) and on Mount Rainier itself to identify areas of present-day hydrothermal activity.

- *Potential field surveys,* including gravity and MT surveys to locate magma and other intrusive bodies, faults, and hydrothermally altered rocks. As part of this work, the existing gravity, magnetic, and MT surveys should be reprocessed to better define the SWCC and its relation to mapped geology.

- *Remote-sensing surveys,* including multiband spectral imaging to map patterns of hydrothermal alteration on the edifice, and SLAR to locate and map small-scale lineaments and faults.

4. Lahar Studies. Special attention should be given to the investigation of lahars, probably the most significant volcanic hazard to life and property in the Puget Lowland area. Detailed mapping, including subsurface mapping, should be carried out to reconstruct the spatial and temporal distributions of these flows and to obtain volumetric estimates for each flow event. This work should be supported by high-precision radiocarbon and dendrochronology analyses to establish flow correlations and chronologies.

5. Edifice stability assessment. Edifice collapse is also a significant hazard to life and property in the Puget Lowland area; it warrants careful study. Research on edifice stability should focus on mapping the distributions of hydrothermally altered rocks, faults, and dikes, which are mechanically weak and prone to failure. The physical properties of hydrothermally altered rocks should be assessed through quantitative mineralogical and mechanical analyses. Research should also focus on the delineation of the hydrothermal system and the process of wallrock alteration, particularly beneath the glaciers that cover the edifice.

4

VOLCANO MONITORING

The previous chapters reviewed the hazards to property and life from Mount Rainier and outlined research needed to provide for better understanding of their nature and frequency of occurrence. This chapter describes monitoring activities that are essential in order to identify anomalous behavior on or around the volcanic edifice that could serve as an early warning of these hazards. Such a monitoring program is necessary to assure that early signs of hazardous activity can be recognized and that its significance for future activity can be evaluated. To be effective, monitoring of the volcano must be carried out at a number of different time and distance scales, using a variety of techniques. Some of the more important techniques are discussed in this chapter.

The use of monitoring techniques to detect anomalous activity presupposes the existence of adequate baseline data that can be used to evaluate such behavior. Ideally, these baseline data would comprise long, continuous series of observations collected at regular intervals during periods when no anomalous activity is occurring. These data series are needed to establish background or average values of behavior that would allow anomalous activity, which would rise above these background values, to be clearly identified.

Adequate baseline data to support monitoring efforts at Mount Rainier are generally lacking. Consequently, an essential aspect of monitoring efforts must involve the acquisition of such data through a program of regular sampling and analysis. The collection of these data could proceed hand in hand with the monitoring activities outlined below.

There are five basic areas for which monitoring and collection of baseline data are important:

1. seismicity;
2. ground deformation;
3. hydrothermal activity;
4. changes in surface appearance of the volcanic edifice; and
5. stream- and debris-flow detection.

The monitoring techniques discussed in this chapter fall into two groups: continuous monitoring, for which measurements are made at regular intervals of seconds to minutes, and intermittent monitoring, for which measurements or observations are made at irregular intervals of days to years. These techniques, developed and tested over the past several decades at active volcanoes around the world, are summarized in Table 4.1 and discussed in the following sections.

Seismicity Monitoring

Seismic monitoring of Mount Rainier can be used to detect the movement of magma beneath or within the edifice that could signal an imminent volcanic eruption. Seismic monitoring can also be used to detect the movement of glaciers on the volcanic edifice or movement of the edifice itself, which could signal impending glacier outburst floods, rockfalls, or slope failures.

Since 1962, one seismometer has operated on Mount Rainier as part of a worldwide seismic station network (WWSSN station LON at Longmire, Washington). Three additional seismometers currently operate within Mount Rainier National Park, and three others in the immediate vicinity function as part of the Washington Regional Seismograph Network operated by the University of Washington. This group of seven instruments (Figure 4.1) is capable of locating an earthquake of magnitude 1.0 or larger within or beneath Mount Rainier.

In the last decade, these seismometers have recorded several thousand events beneath or within the volcano. A few hundred events were clearly earthquakes; this number of events makes Mount Rainier the second most seismically active volcano in the Cascade Range north of Califor-

FIGURE 4.1 Current seismic stations (circles) and earthquakes (squares) with magnitudes > 3 over the past 25 years.

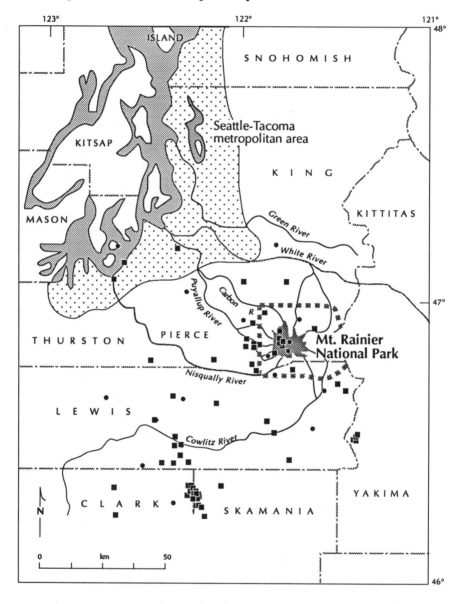

TABLE 4.1 Examples of Monitoring Techniques Applicable to
Mount Rainier

Continuous (telemetered)	Intermittent
Seismic network	Visual observations
Stream gages	Photography
Acoustic or vibration network	Infrared survey
Tiltmeters	GPS campaign
Continuous GPS	EDM survey
Slope-stability meters	Radar interferometry
Gas monitors	Radar survey of ice thickness
	Gas chemistry
	Stream chemistry

nia, after Mount St. Helens. The other events were probably caused by the
movement of glaciers or by rockfalls on the flanks of the volcano. Earthquakes currently are centered beneath Mount Rainier at depths of 5 km or
less as well as away from the mountain at depths of 10 to 20 km. A broad
zone of earthquakes occurs about 15 km west of Mount Rainier. Since
1972, the largest earthquake in that area was a magnitude-4.1 earthquake
that occurred on July 29, 1988 (University of Washington Geophysics
Program, 1993). The relation between this zone of nearby earthquakes and
volcanic activity at Mount Rainier is unknown.

With the present array of seismometers, it is difficult to distinguish
between surface seismic events caused by glacier movements on the edifice
and shallow earthquakes beneath Mount Rainier (Weaver, 1976; Malone
and others, 1991). This distinction is important, because shallow earthquakes might reflect the movement of magma to shallow levels in the
volcano and could signal an impending eruption. On the other hand, an

increase or change in surface seismic events could presage a glacier out-burst flood or rockfall, either of which could produce a lahar. Clearer distinction and more reliable detection of shallow earthquakes and surface seismic events could be made if two or three additional seismometers were placed in operation on the upper slopes of the volcanic edifice, especially if these seismometers were three-component instruments capable of detecting ground motion with high resolution.

Monitoring of Ground Deformation

Two common precursors of volcanic eruptions are uplift and lateral distension of the ground surface caused by upward movement of magma beneath and into the volcano. Such tumescence may involve a portion of the volcano, the entire volcano, or a broad region around it. Similarly, a common precursor to large landslides is the slow creep or slumping of a portion of the volcanic edifice. The detection and measurement of these movements using the techniques described below could provide days to months of warning of impending eruptions or edifice failures. Such movements may be no larger than a few centimeters in magnitude; consequently, high measurement precision is required to detect them.

Topographic and road-building surveys conducted around Mount Rainier in the early 1900s provided the first geodetic control of the region. No systematic pattern of ground movement has yet been recognized based on comparisons with these early surveys. The first surveys specifically designed to measure ground deformation of Mount Rainier were made in 1982 (Dzurisin and others, 1983; Chadwick and others, 1985; see Figure 4.2). Measurements to determine horizontal positions of about a dozen stations around the volcano have been repeated three times and indicate no horizontal ground movement within the precision of the measurements except for one clearly unstable benchmark (Iwatsubo and Swanson, 1992).

The elevation of the summit of Mount Rainier was measured in 1988 using GPS (DeLoach, 1989), which utilizes satellites to determine ground positions to within a few centimeters over distances of a few hundred kilometers. This system is useful for the study of active volcanoes,

FIGURE 4.2 Geodetic network at Mount Rainier. Benchmarks are shown by numbers and names (from Iwatsubo and Swanson, 1992, Figure 10.6).

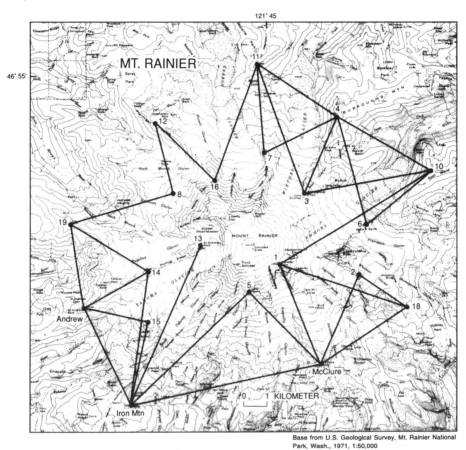

Base from U.S. Geological Survey, Mt. Rainier National
Park, Wash., 1971, 1:50,000

because measurements can be made in remote or rugged terrain and in poor weather conditions. The survey data can, in some configurations, be telemetered to a central point for recording and interpretation.

Good geodetic control of Mount Rainier could be obtained through real-time continuous GPS monitoring or frequent GPS surveys of the existing geodetic stations in the region and by the establishment of additional stations at higher elevations on the edifice. Frequent measurements of many points are needed to distinguish between changes in elevation due to magma movement and changes due to the seasonal accumulation and melting of snow on the edifice.

To be most effective in detecting ground deformation, the local network of GPS stations on Mount Rainier must be integrated into the regional network that extends about 100 km from the volcano. An active volcano is part of a larger system of the Earth's crust in which regional contraction or extension may promote the movement of deep-seated magma into the edifice. Only since the advent of GPS has it been practical to search for a relation between regional and local activity around active volcanoes.

Measurements of strain changes on time scales of both years (from geodetic surveys) and seconds (from seismometers) are needed, both on the volcano and in surrounding areas. Strain buildup may persist for weeks prior to an eruption, and creep presaging massive landsliding may last for days or even months. Continuous monitoring needed to track such rates of deformation requires tiltmeters, strainmeters, dilatometers, and GPS. Some of this instrumentation (e.g., seismometers, geodetic benchmarks) is currently in place, but additional instrumentation needs to be developed and installed to provide continuous real-time monitoring of the edifice and surrounding areas. Of particular importance is the development and installation of instrumentation to continuously monitor the stability of the flanks of the volcano, because flank failure is a potentially significant hazard to life and property, as discussed in previous chapters.

Deformation of the edifice might also be measurable using a new technique, radar interferometry, that is under development as part of the Earth Observing System of the National Aeronautics and Space Administration. This technique, which utilizes satellite radar to determine differential vertical movements as small as 1 cm, has been used to detect and measure the displacement field associated with the 1992 earthquake in Landers, California (Massonnet and others, 1993). This satellite-radar system could be used at Mount Rainier as a complement to GPS measurements and to foster development of the technique. This radar system may also be useful for monitoring ice cover on the edifice, as described in a later subsection.

Monitoring of Hydrothermal Activity

Another potential precursor of volcanic activity is a change in the composition or rate of emission of gases and hydrothermal fluids from the volcano. The detection of such changes is facilitated by long-term records of baseline measurements acquired through a program of regular sampling and analysis. At present, such measurements do not exist, but they will be required for evaluating the nature, magnitude, and significance of future changes unless the changes are large.

Thermal springs, fumaroles, and warm ground are found on and near Mount Rainier (Frank, in press), primarily in the summit area (see Figure 3.3). Studies of these hydrothermal features have focused on measurements of temperature and fluid composition, utilizing samples collected at infrequent intervals. The fumaroles in the summit area emit mostly warm air with trace amounts of hydrogen sulfide, rather than gases of dominantly magmatic origin (Frank, in press).

Future activity at Mount Rainier will likely include the emission of a visible gas plume, which would probably appear as a light gray to brown haze extending downwind from the volcano. This plume will probably contain gases such as sulfur dioxide and hydrogen sulfide, which are usually detectable by odor, as well as carbon dioxide, halogens, and other species. The remote measurement of such plume gases would be invaluable for determining the presence of shallow magma bodies that could feed volcanic eruptions. Correlation spectrometry (COSPEC) and infrared spectrophotometry (MIRAN) measurements from the ground or from airplanes or even satellites could be used to detect and measure sulfur dioxide and carbon dioxide in the plume.

Measurement of stream-water chemistry could be used to detect subglacial hydrothermal or fumarolic activity that could produce catastrophic melting of snow and ice on the volcanic edifice. Such measurements would be particularly important when changes in glacial morphology were observed or when high rates of glacier-related seismicity were detected. Chemical measurements must be obtained at sufficiently frequent intervals to determine "normal" background values. At a minimum, sampling of fluids during each season of the year would contribute to an understanding of the influences of snowpack and rainfall on fumarole gas, condensate chemistry and temperature, stream chemistry, and stream discharge. Such monitoring proved useful during the thermal event at Mount Baker during the mid-1970s (Frank and others, 1977).

Monitoring Changes in Surface Appearance

The distribution of snow and ice cover on Mount Rainier is controlled by seasonal and climatic changes and, on a local scale, by heat flow from the volcano. Longer-term (decadal) climatic changes mainly control the volume of glacier ice on the edifice (Figure 4.3). Shorter-term seasonal changes mainly affect the annual accumulation and melting of the snowpack. Very short-term changes (much less than 1 year), due to increases in hydrothermal or fumarolic activity or transient heating events, could affect both the volume of glacier ice and the distribution of the winter snowpack. A short-term, nonseasonal change in snowpack or ice volume might signal increased thermal activity that could produce catastrophic outburst floods, debris flows, or, possibly, eruption.

Several techniques could be usefully employed to monitor changes in snow or ice cover on the volcanic edifice:

- *Visual Observation.* The staff of Mount Rainier National Park is familiar with the volcano and is therefore likely to recognize subtle nonseasonal changes in the distribution of snow and ice that might

FIGURE 4.3 Oblique photograph of the south flank of Mount Rainier showing the permanent cover of snow and ice. Nisqually glacier, one of 25 named glaciers on the edifice, is shown in the foreground. (Courtesy of David Hirst, USGS.)

not be apparent to visiting scientists or the general public. Observations of changes in glacier or snow morphology by park staff and frequent visitors need to be reported to scientists responsible for monitoring the volcano, both to alert them to the observations as well as to help interpret the records and data from other monitoring activities.

- *Photogrammetry.* Existing oblique air photos of Mount Rainier and its glaciers date back to the 1930s (Frank, 1985). Vertical air photos are also available but are of limited use for mapping purposes because of inadequate control of ground points. High-quality aerial photography is needed for the volcano, including surveys of the snow and ice cover. These images can be used to monitor the volcano and to assist with detailed geologic mapping, identification of potential landslide areas, and preparation of detailed elevation charts.

- *Infrared Heat Emission.* Since the 1960s, hot spots around the summit area have been mapped occasionally using infrared sensors (Moxham and others, 1965; Frank, 1985). These hot spots define the margin of the main summit crater. Continued thermal infrared imagery of this area and of selected areas on the slopes of the volcanic edifice would provide the baseline information necessary for revealing a significant change in the pattern of heat flow related to a change in the pathways of warm water through the volcano. The appearance of new hot spots on the slopes might be the only clear indicator that a rocky prominence is being weakened by percolating groundwater and is thus susceptible to failure.

- *Radar Imagery.* The use of radar to measure ice thickness has been successful on permanent ice sheets but may not always work well in temperate settings such as Mount Rainier. Glacial ice on Mount Rainier generally contains varying but significant amounts of liquid water, which would limit the penetration of radar signals. However, many of the glaciers on Mount Rainier are thin enough to allow radar signals to penetrate to the underlying bedrock and thus could be used in repeat surveys to detect and measure gross thickness changes. Driedger and Kennard (1986) successfully used radar imagery to determine ice volumes at Mount Rainier and three other Cascade volcanoes.

As with the previously discussed monitoring activities, the key to success in detecting changes on the surface of the volcanic edifice lies in frequent observations combined with the acquisition of baseline data through a program of regular sampling and analysis.

Detection of Stream Flow and Debris Flows

The rapid melting of even a small portion of the extensive cover of snow and ice on the edifice could generate large floods, which could entrain a large amount of volcanic debris to produce lahars in streams draining the volcano. Catastrophic floods and lahars would doubtless be generated by a major volcanic eruption of the edifice. Major floods and lahars also could be generated by transient thermal events, which might have no obvious surface expression. Such events could occur without warning. Consequently, it is essential to monitor the edifice for floods and debris flows after they have formed and are moving downslope.

Several hours generally pass between the initiation of a major lahar from Mount Rainier and runout of the lahar onto the lower slopes and floodplains surrounding the volcano (Scott and others, 1992). The vibrations set up by moving lahars can be detected by seismometers and acoustic sensors. Modified seismic systems could provide inexpensive yet robust debris-flow detectors and could be linked to the existing network of seismometers designed to record earthquakes. Such systems have provided warning at Redoubt Volcano, Alaska, and Mount Pinatubo, Philippines (Hadley and Lahusen, 1991).

The streams and rivers around Mount Rainier are fed by water from rainstorms as well as from melting of the ice and snow on the edifice. The addition of more weather stations and the operation of a network of stream gages to record changes in water discharge are first steps in revealing the complex water budget of Mount Rainier. Stream-flow monitors high on the volcano can also detect the sudden changes in water level that may precede a lahar.

Recommendations

A program of volcano monitoring should be established at Mount Rainier to identify anomalous activity that could serve as an early warning of the occurrence of volcanic hazards such as eruptions, edifice collapses, and lahars. This monitoring program should include plans for the collection of adequate baseline data—long, continuous series of observations acquired at regular intervals when no anomalous activity is occurring—in order to provide a background of values with which to contrast anomalous behavior. Efforts to collect these baseline data should proceed hand in hand with the regular monitoring activities.

Monitoring should utilize the following techniques:

1. Seismic monitoring, to detect the movement of magma, glaciers, and rock on or beneath the volcano. The present network of seismometers (Figure 4.1) is sufficient to detect small-magnitude earthquakes associated with magma movement, glacier movement, and rockfall, but it is inadequate to locate these events in space precisely. This network should be upgraded with additional three-component seismometers located on the slopes of the volcanic edifice. A minimum of two seismometers should be added to this network.

2. Ground-deformation monitoring, to detect magma movement and edifice creep. The present network of geodetic stations in the region (Figure 4.2) should be expanded, with additional stations established at higher elevations on the edifice; this local network should be integrated into the regional network surrounding the volcano. This expanded network should be monitored using real-time, continuous GPS, or resurveyed at frequent intervals using GPS, so that seasonal changes in shape due to the annual accumulation and melting of snow and ice can be detected, characterized, and distinguished from hazard-related ground movement. Tiltmeters, strainmeters, and dilatometers should be installed on and around the volcanic edifice to monitor creep and deformation. Efforts to monitor deformation should include the development of techniques to monitor the stability of the volcano flanks. Satellite radar interferometry is recommended as potentially useful for such monitoring.

3. *Monitoring hydrothermal activity, to detect changes in the composition or rates of emission of gases and fluids from the edifice.* A program of fluid and gas sampling should be initiated to detect changes in the hydrothermal system within the volcanic edifice. This program should include monitoring for gas plumes if the volcano shows signs of anomalous behavior, such as increased seismicity or significant ground deformation. Samples should be collected and analyzed on a periodic basis from streams, thermal springs, and fumaroles in order to establish a baseline against which to recognize future changes. For geochemical measurements of springs and streams, data analysis should take into account the effects of precipitation and stream discharge.

4. *Monitoring changes in surface appearance, to detect changes in snow and ice cover on the volcano.* A program to monitor changes in snow and ice cover on the volcano should include visual observation, photogrammetry, infrared heat emission, and radar imagery. Staff at Mount Rainier National Park, who live and work on the volcano and are most likely to recognize anomalous changes in the ice and snow cover, should be enlisted to provide regular visual observations of the volcanic edifice. These visual observations should be augmented with high-resolution, vertical photogrammetry to map the distribution of snow and ice cover on the volcano. This imagery should be supplemented with radar imagery to map the thickness of the snow and ice cover on the edifice. Infrared thermal emission images should also be collected to monitor the distribution of hot spots on the volcano.

5. *Stream monitoring, to detect floods and debris flows.* A network of sensors should be installed in the major drainages on the edifice to detect the formation and movement of debris flows after they have formed and are moving downslope toward populated areas. These sensors should be tied into the existing seismic network. Additional weather stations and stream gages should be installed in the park to interpret the data from this sensor network and to establish the water budget for the volcano.

5
MITIGATION: COEXISTING WITH MOUNT RAINIER

About 3.5 million people live and work in proximity to Mount Rainier. Many residents of this region may be unaware of the hazards posed by the volcano. Wholesale, permanent evacuation of the region around the volcano (large parts of Pierce, King, Lewis, and Cowlitz counties; see Figure 2.1) would be necessary to completely eliminate risk to life and property from the volcano. Obviously such an approach is unrealistic and unworkable. The communities in the region must seek ways to reduce risk to life and property from volcanic hazards while maintaining the strong economic base that derives in part from the desire of people to live, work, and play around the volcano. This chapter addresses important mitigation measures that can be taken to reduce the risk from the volcanic hazards described in previous chapters of this report.

An effective risk-mitigation strategy can be undertaken only as part of a comprehensive strategy to understand the volcano. Effective risk mitigation requires that (1) hazards are well understood; (2) they can be recognized before they reach a critical level; (3) warning of their occurrence can be communicated quickly, clearly, and accurately to public officials; and (4) public officials will understand the significance of such warnings and will initiate appropriate mitigative measures. The present chapter addresses the communication of information and warnings on hazards to responsible authorities and the general public in order to mitigate risk through planning and implementation of risk-reducing measures.

Communication

Effective mitigation requires communication at several different levels among the many groups who live and work near the volcano:

- within the scientific community;
- between scientists and responsible authorities;
- between scientists and the general public; and
- between responsible authorities and the public.

Broadly speaking, this communication can be divided into two categories: communication during times of volcanic quiescence and communication when signs of unrest have been detected or when an actual eruption or related event has begun.

During volcanic quiescence, such as exists at the present time, scientists working on Mount Rainier, as well as those contemplating such work, generally communicate with each other through publication in scientific journals and presentations at scientific meetings. There is a need to make this literature available in one place to working scientists and nonspecialists in a more timely fashion. There is also a need to provide information about current and planned projects, which is normally not communicated effectively through the scientific literature.

Scientists and nonscientists alike would benefit from the establishment of a Mount Rainier Hazards Information Network accessible through the Internet. Such a network would be useful for sharing information about current and planned research, published reports on the volcano, hazards identification, and risk-mitigation projects. To be most useful, information and data could be made available in both text and Geographic Information Systems (GIS) formats. The same network could be linked with county planners and emergency-service planners to promote the application of research findings. The results of much of the research are likely to be communicated in unpublished (and unrefereed) reports; consequently, the legal and ethical implications of this network for local decision making must be considered.

The present quiescence also provides an opportunity to educate responsible authorities and the general public about the nature of expected hazards at Mount Rainier and how their effects might be mitigated. Such communication could take place in a number of ways. For example, scientists could work with educators to prepare high-impact educational materials about volcanic hazards and risk-mitigation options. These materials could include videotape productions and short, well-illustrated pamphlets or brochures. The videotapes could be aired on local television; the printed materials could be designed as an insert in local newspapers and for general distribution when precursory activity is detected. These materials could also be distributed during presentations by scientists at schools and public meetings. Because many Park Service employees have developed skills for communication with the public and are in regular contact with many thousands of park visitors, they can play an important role in the design, preparation, and distribution of much of this material. Scientists and Park Service staff could also work together to develop educational displays on volcanic hazards and emergency response for visitors to Mount Rainier National Park. The effectiveness of these educational materials needs to be evaluated periodically in order to maximize their usefulness.

Communication becomes even more important when precursory activity is detected or during an actual event such as an eruption. The Mount Rainier Hazards Information Network discussed above could serve as a real-time clearinghouse for information, helping to lessen the redundancy that can occur when large numbers of scientists converge on an active volcano. The same network could be used to increase communication between scientists and responsible authorities who provide hazard warnings to the public. In this context, responsible authorities include the National Park Service; local and state governments; and local, state, and federal emergency management agencies.

This communication needs to be part of a comprehensive emergency response plan for scientists involved in volcano monitoring to communicate with responsible authorities when conditions warrant, such as when precursory activity is detected or during actual events. To be effective, this emergency response plan must be integrated into on-going emergency planning by state and local authorities and must contain the following elements:

- well-defined classifications of possible precursory behavior (for example, elevated levels of earthquake activity or edifice creep) detected by monitoring networks cross-referenced to possible hazards, as well as a procedure for regularly updating and communicating this list to emergency-management personnel;
- a shorthand code for easily communicating the level of concern implied by the precursory behavior, such as the color-coded warning levels currently used at the Alaska Volcano Observatory; and
- contact lists of emergency-management personnel and scientists involved in monitoring, as well as procedures for keeping these lists up to date. For example, Norris (1991) provides a summary of monitoring information, involved scientists, and pertinent contacts in the National Parks, National Forests, and other land holdings in the Cascades; the Cascades Volcano Observatory and the University of Washington maintain lists of contacts for responding to events at Mount St. Helens.

Communication between responsible authorities and the public during a crisis is essential to provide timely and accurate warnings and information about volcanic hazards. This communication needs to be planned in advance of an actual emergency, and its effectiveness needs to be periodically tested and evaluated through table-top exercises and simulations. Such exercises need to involve the news media; local and state governments and emergency-management agencies and the U.S. Federal Emergency Management Agency (FEMA); the National Park Service and U.S. Forest Service; Washington Department of Natural Resources; and scientists involved with volcano monitoring.

The effectiveness of communication between scientists and the general public could be increased through the use of "scientist-spokespersons" who have had training or experience or both working with the news media. This is an important communication link, because timely and accurate information is essential for public understanding and cooperation during a hazardous event. Fiske (1984, p. 176) suggests that the role of such "information officers" is to:

> work closely with the chief scientist to ensure that a single and complete stream of information is made available. This person should not suppress scientific disagreements that might exist between working scientists but should express them freely in terms of overall scientific understanding. It is common for scientists to disagree, and such disagreements should be reported publicly in a balanced and nonpersonalized way.

An information scientist at the Cascades Volcano Observatory has provided information to the media for activity at Mount St. Helens since 1980.

Inevitably, however, individual scientists, many of whom have had little experience communicating with the media, will be contacted about an exclusive story or "angle." These scientists need to cooperate with the media as much as possible while avoiding the temptation to speculate on the outcome of events that cannot be accurately predicted. The effectiveness of media communications with the public will be greatly increased if the public's understanding of volcano hazards has been already improved by the various pre-eruption educational activities mentioned above.

Planning and Implementation

The reduction of risk from volcanic hazards requires planning and implementation of effective risk-reducing measures. To be successful, scientists, government, business, and other citizens need to be involved in planning, implementing, and periodically evaluating and adjusting specific measures as needed. Several measures, if implemented, would significantly reduce risk from volcanic hazards to people and property in the region:

- *Risk analyses*, to assess risks to populations and businesses from specific volcanic hazards such as tephra and debris flows.

- *Land use planning and regulation*, to encourage, and where appropriate require, uses and construction practices that are appropriate to the degree of risk in areas potentially affected by eruptions or lahars. For example, open-space uses could be mandated for areas with extremely high hazards.

- *Emergency planning*, to deal with specific volcanic hazards, including planning for evacuations and temporary housing before and during eruptions or lahars.

- *Engineering solutions* for specific hazards; for example, the construction of sediment traps or diversion structures to protect populated areas from lahars.

- *Economic incentives* for business and citizens to reduce their risk from specific hazards; for example, reducing risks from debris flows by offering lower casualty insurance rates for structures built on high ground.

An important contribution of geoscientists in these efforts is the identification of areas at risk through the development of *hazard maps* (for example, Crandell, 1973; Scott and others, 1992; see Figure 5.1). Hazard maps are spatial representations of areas at risk from lava flows, debris flows, tephra falls, pyroclastic flows, lateral blasts, glacier outburst floods, massive slope failure, and similar events. Data for these maps are obtained from predictive studies or from past behavior of the volcano as determined from field studies, such as those discussed in Chapter 3 of this report. The maps can be drawn for events of any magnitude and can be presented in a probabalistic context. Hazard maps produced in GIS format using Digital Terrain Model (DTM) frameworks are particularly useful, because they lend themselves to the production of sketches, cross-sections, and sequential diagrams illustrating the development and consequences of hazards.

Hazard maps, in turn, can be used to prepare *risk maps* or *vulnerability indices*. Such maps and indices are developed by combining hazard maps with demographic and geographic data, for example: population density; the locations of critical facilities (e.g., hospitals); transportation routes; and residential, industrial, agricultural infrastructure. This superposition would show where populations, businesses, and critical facilities are

FIGURE 5.1 Hazard map showing estimated risk (low, medium, high) from tephra eruptions and debris flows at Mount Rainier (modified from Crandell, 1973). The risk estimates shown on this map are based on work prior to 1973 do not reflect data collected during the past two decades (e.g., Scott and others, 1992). The map is used here for illustration purposes only.

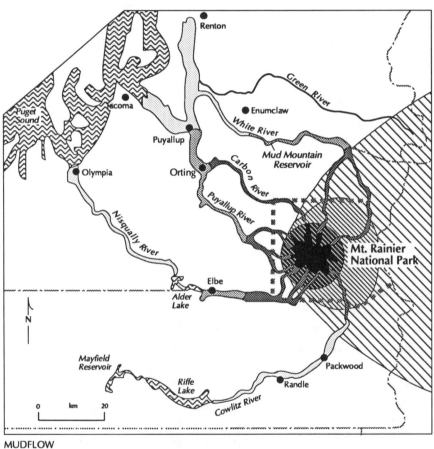

MUDFLOW

▨ High Risk
Valley-floor areas in which there could be a high degree of danger from relatively small but frequent mudflows and floods

▨ Moderate Risk
Valley-floor areas which could be covered by a mudflow as large as the Electron Mudflow, in valleys in which mudflows have been relatively frequent

▢ Low Risk
Valley-floor areas which could be covered by a mudflow as large as the Electron Mudflow, in valleys in which mudflows have been relatively infrequent, and areas in the other valleys which might be subject to flooding

TEPHRA

▧ High Risk
Areas in which there could be a high degree of danger to human life and property from asphyxiation, falling rock fragments, and accumulation of tephra

▨ Moderate Risk
Areas in which there could be a low degree of danger to human life and a moderate degree of danger to property from falling rock fragments and accumulation of tephra

◸ Low Risk
Areas in which there could be a low risk of danger to property from accumulation of tephra, but virtually no direct risk to human life

especially vulnerable to hazards. Such maps could be used to guide the relocation or "hardening" of critical facilities, such as hospitals, utilities, and pipelines. They would also serve as a zoning tool to guide future development and could be used by the insurance industry to establish casualty-protection rates.

Geoscientists can play a significant role in the development of risk maps and vulnerability indices by working cooperatively with planners, engineers, social scientists, and legal professionals to ensure that hazard maps contain appropriate data presented in formats understandable for hazard-mitigation efforts. For example, the preparation of hazard maps using basemaps and GIS formats currently employed by state, county, and city planning agencies would increase their usefulness for land use planning and regulation and vulnerability analyses. Volcano scientists need to work closely with city planning agencies, engineering/public works departments, transportation agencies, and emergency-management coordinators to anticipate the consequences of volcanic hazard assessments for people, real estate, utilities, communications, and transportation activities.

Once basic information about hazards, risks, and actual or potential risk-mitigating measures is in the hands of the public and responsible authorities, debate can begin about which specific measures to adopt. Analyses of potential risk-mitigation measures need to address both economic costs and benefits as well as the possible costs of inaction. These analyses must be put forward for public debate, and measures, once adopted and implemented, need to be periodically tested and fine-tuned.

As part of this process, it is important to document how information about Mount Rainier hazards is presently incorporated, and is planned to be incorporated, into the State of Washington's Growth Management Act of 1990 (Appendix A), emergency plans, and engineering plans. For example, volcanic hazards information is presently being considered in Pierce County because of requirements of the Growth Management Act. Careful tracking of how that information is (or is not) translated into actual mitigative measures is needed to transfer successful ideas to other jurisdictions, or to spot jurisdictions for which more information, more aggressive application, or alternative measures are needed.

Natural scientists traditionally participate in this process up to the point of public debate, providing information about hazards and, occasionally, about their effects on people. Social scientists, on the other hand, rarely become involved in hazard studies. The U.S. Geodynamics Committee believes that more effective risk-mitigation measures can be designed and implemented if natural and social scientists work together through the entire process. Similarly, social scientists, geoscientists, planners, engineers, decision makers, and the general public must work together, from hazard assessment through implementation of risk-mitigation measures, if the populations around Mount Rainier are to coexist in reasonable safety with the volcano.

Social Aspects of Mitigation

In general, there are reasonably good land use, building, and development practices for reducing risks from natural hazards. Less well understood is how to motivate the implementation of these measures. A major gap exists between theory and practice with respect to risk reduction and mitigation for natural hazards.

Societal impacts of a major eruption or debris flow are potentially enormous in the aggregate but are likely to be quite varied (Cullen, 1978; Cullen Tanaka, 1983). Some geographic areas, sectors of the economy, and population groups would be affected more than others. Planners and scientists tend to think about aggregate impacts of hazards, whereas the effects of impacts tend to vary across populations. This distinction has important implications for risk assessments, which need to focus on vulnerabilities of different populations, geographic areas, and sectors of the economy.

Mitigation efforts also have potentially significant social impacts. For example, the implementation of development restrictions in areas judged to be of high risk could lower land values and cause economic hardships for landowners. These potential impacts need to be understood to develop socially and politically acceptable mitigation strategies.

The impact of a given event is determined by the physical properties of the hazard, the "social fabric" of the region, and the resiliency of its population. Some sectors of the population, such as elderly and low-income groups, are less resilient in responding to and recovering from disasters. Similarly, some communities are better able to recover from disasters. Because of differences in local economies and political jurisdictions, there is considerable variation in the application of land use and development measures to reduce potential volcanic hazards. Risk assessments need to consider these factors in targeting efforts to reduce losses.

For example, mountain-dependent communities, including those that act as "gateways" to the National Park, would likely suffer great loss of property and life in the event of a major eruption or debris flow. Further, these communities could have a more difficult time recovering from such events because their economies are dependent on tourism, which, in the short term, would probably decline following a major event. (Later, tourism would probably increase, as it has at Mount St. Helens.)

Social response to a given hazardous event is influenced by several factors. These include the nature of the event (e.g., eruption versus debris flow), timing (e.g., summer versus winter, day versus night), and emergency preparedness. Another important factor for Mount Rainier is the potential for multiple hazards from a single event, for example, a large earthquake that collapses the edifice to produce an eruption and debris flow. Such multiple hazards greatly complicate efforts to predict social responses. Nevertheless, such hazards are a real possibility and therefore need to be considered explicitly in planning.

Natural hazards are issues that normally show up on government agendas only after they occur. The challenge is to put mitigation measures into effect in the face of official indifference or inaction. A particularly effective approach is to include risk-mitigation plans in related pieces of legislation. A pertinent example of such legislation is the Growth Management Act (Appendix A), which requires state and local government agencies to include consideration of natural hazards in planning future development.

The issues of risk perception, hazard mitigation, and policy design as they relate to Mount Rainier are appropriate topics for social science investigation. Relevant policy questions concern vulnerability, liability, and cost in addressing potential hazards. There is a limited amount of social science research concerning volcanic eruptions in the United States, principally Mount St. Helens (Sheets and Grayson, 1979; Johnson and Jarvis, 1980; Sorensen, 1981; Warrick and others, 1981; Dillman and others, 1982; Kartez, 1982; Leik and others, 1982; Perry and Greene, 1983; Saarinen and Sell, 1985; Buist and Bernstein, 1986; Perry and Lindell, 1986), but there is a strong foundation of relevant research (Lachman and Bonk, 1960; Marts, 1978; Blong, 1984; May, 1985; Mader and Blair, 1987; Perkins and May, 1987) concerning other volcanoes and natural hazards that serves as a basis for framing this discussion. Clearly, additional applied research is needed in this area.

Recommendations

Mitigation of risk is an important component of the Mount Rainier Decade Volcano Demonstration Project and can be successfully executed only within the context of a comprehensive strategy to understand the volcano and its various hazards. The success of mitigation efforts requires that the hazards themselves are well understood; that they can be recognized before they reach a critical level; that warning of their occurrence can be communicated clearly, accurately, and quickly to public officials; and that public officials will act to put the appropriate risk-mitigating measures into operation. The important elements of an effective mitigation program are these:

1. Communication is essential among the many groups that live and work around the volcano, including:

- *Within the scientific community.* Communication within the scientific community is necessary to coordinate and disseminate research on the volcano. To this end, the U.S. Geodynamics Committee recommends the establishment of a Mount Rainier Hazards

Information Network on the Internet to disseminate past, current, and planned research and information on mitigation measures. This network should be accessible to researchers, National Park staff, county planners, and emergency-services planners.

- *Between scientists and responsible authorities.* An emergency-response plan should be developed so that scientists involved in monitoring can provide responsible authorities with accurate and timely warnings of impending hazards and can keep officials informed during such events. This plan should be integrated into ongoing emergency planning by state and local authorities.

- *Between scientists and the public.* Scientists should work with educators and National Park Service staff in times of quiescence to inform the general public about the nature of volcanic hazards, people and property at risk, and options for risk reduction through presentations at schools and public meetings, by the preparation and distribution of high-impact educational materials, and by the development of displays on volcanic hazards and emergency response for visitors to Mount Rainier National Park.

- *Between responsible authorities and the public.* Responsible authorities should develop plans for communicating timely and accurate information and warnings about volcanic hazards to the public. The effectiveness of such plans should be periodically tested and evaluated through table-top exercises and simulations; this testing should involve the news media, local and state governments, emergency-management agencies, the National Park Service, the U.S. Forest Service, Washington State Department of Natural Resources, and scientists involved with volcano monitoring.

2. Planning and implementation of risk mitigation measures should involve scientists, government, business, and citizens and should be coordinated and, where appropriate, integrated with other planning activities in the region. Several measures, including the following, should be considered for implementation in order to significantly reduce risk from volcanic hazards to people and property:

- analyses to identify regions and populations at risk;
- land use planning to encourage appropriate use of high-risk areas;
- engineering solutions to mitigate risks, where possible, from specific volcanic hazards; and
- economic incentives to encourage business and citizens to reduce risk from specific hazards.

Costs and benefits of specific mitigation measures should be put forward for public debate and, once implemented, these measures should be periodically tested and adjusted as necessary.

Societal impacts of a major eruption or debris flow would likely be enormous in aggregate, but these effects would probably be spread unevenly across different communities and population groups. In planning and implementing specific mitigation measures, planners should focus on the vulnerabilities of different sectors of the population, region, and economy.

There should be a critical evaluation of the existing social science literature as to its specific relevance to Mount Rainier. Future research on the social consequences of a Mount Rainier eruption should include development of a generic response model that could serve as a "check list" for understanding hazard response and mitigation. There should be efforts to develop a more refined response model for Mount Rainier with gaps in knowledge identified as a research agenda.

6
IMPLEMENTATION

The preceding chapters have outlined a science plan for studying and monitoring Mount Rainier as a Decade Volcano Demonstration Project. Communication and coordination among the many scientists and organizations who will be involved in this project are essential to its implementation in an effective and cost-efficient manner.

Implementation of the Mount Rainier Decade Volcano Demonstration Project is the responsibility of the scientific community, which needs to develop a plan to carry the project forward. This community includes:

- state, federal, academic, and industry researchers who study volcanoes, hazards, and risk mitigation;
- government agencies with relevant extramural research funding programs, such as the National Science Foundation; and
- government agencies with responsibilities for volcano and hazards research, public-lands management, and public safety, including the Washington State Department of Natural Resources, U.S. Geological Survey, National Park Service, Forest Service, and federal and state emergency management agencies.

The implementation plan should provide guidance on:

- priorities for research and monitoring activities based on scientific significance and value to risk-mitigation efforts;
- funding for research and monitoring activities deemed to be of high priority;

- mechanisms for coordinating the efforts of scientists to avoid unnecessary duplication, particularly in the use of instrumentation or collection of samples from wilderness and other environmentally sensitive areas with limited access; and
- mechanisms for balancing the needs of scientists for access, samples, and data with the needs of federal and state agencies to fulfill their research, public safety, and land-management missions.

To be effective, monitoring efforts will require continuity in funding, management, personnel, and facilities that can best be provided by federal and state agencies with responsibilities for volcano and hazards research, such as the U.S. Geological Survey and Washington State Department of Natural Resources. Nongovernment scientists should be encouraged to participate in monitoring activities in both data collection and advisory capacities, and the scientific community should have free and immediate access to monitoring data. A possible mechanism of access would be the Mount Rainier Hazards Information Network, which is discussed elsewhere in this report.

Many of the research, monitoring, and mitigation activities described in this report will require access to Mount Rainier National Park and surrounding Forest Service and private lands for field work, sample collection, and the installation and operation of scientific instruments and telemetry equipment. Much of this land is environmentally sensitive and is designated as wilderness area. Research and monitoring activities must be designed to minimize impacts to the environment. Consultation with Park Service and Forest Service staff for work on federal lands (see, for example, Appendix B) and with state personnel for work on private lands must begin at the design stage of all projects in order to assure compliance with existing regulations.

Park Service and Forest Service staff can make significant contributions to the research and monitoring efforts outlined in this report. They are in a position to notice subtle changes in the volcano that might not be apparent to visiting scientists or the general public. They can make regular visual observations of snow, ice, and rocks on the volcanic edifice; assist with the collection of data; and, where appropriate, assist with inspections and routine maintenance of instrumentation. Cooperation between researchers and Park Service and Forest Service staff is essential to the successful implementation of this project.

REFERENCES

Atwater, B. F., 1987, Evidence for great Holocene earthquakes along the outer coast of Washington state: Science, v. 236, p. 942–944.

Atwater, B. F., and Moore, A. L., 1992, A tsunami about 1000 years ago in Puget Sound, Washington: Science, v. 258, p. 1614–1617.

Bacon, C. R., 1983, Eruptive history of Mount Mazama and Crater Lake caldera, Cascade Range, U.S.A.: Journal of Volcanology and Geothermal Research, v. 18, p. 57–115.

Blackwell, D. D., Steele, J. L., Kelley, S., and Korosec, M. S., 1990, Heat flow in the state of Washington and thermal conditions in the Cascade Range: Journal of Geophysical Research, v. 95, p. 19,495–19,516.

Blong, R. J., 1984, Volcanic Hazards: A Sourcebook on the Effects of Eruptions: Sydney, Academic Press, 424 pp.

Brandon, M. T., and Calderwood, A. R., 1990, High–pressure metamorphism and uplift of the Olympic subduction complex: Geology, v. 18, p. 1252–1255.

Brunstad, K. A., Hammond, P. E., and Hooper, P. R., 1992, Caldera-fill mechanisms at Fifes Peaks caldera, Cascade Range, southern Washington [abs.]: EOS, Transactions of the American Geophysical Union, v. 73 (43), p. 612.

Bucknam, R. C., Hemphill-Haley, E., and Leopold, E. B., 1992, Abrupt uplift within the past 1700 years at southern Puget Sound, Washington: Science, v. 258, p. 1611–1614.

Buist, A. S., and Bernstein, R. S. (editors), 1986, Health effects of volcanoes: an approach to evaluating the health effects of an environmental hazard: American Journal of Public Health, Supplement to Volume 76 (March), 90 pp.

Cameron, K. A., and Pringle, P. T., 1987, A detailed chronology of the most recent major eruptive period at Mount Hood, Oregon: Geological Society of America Bulletin, v. 99, p. 845–851.

Carrasco-Núñez, G., Vallance, J. W., and Rose, W. I., Jr., 1993, A voluminous avalanche-induced lahar from Citlaltépetl Volcano, Mexico: implications for hazard assessment: Journal of Volcanology and Geothermal Research, v. 59, p. 35–46.

Chadwick, W. W., Jr., Iwatsubo, E. Y., Swanson, D. A., and Ewert, J. W., 1985, Measurements of slope distances and vertical angles at Mount Baker and Mount Rainier, Washington, Mount Hood and Crater Lake, Oregon, and Mount Shasta and Lassen Peak, California, 1980–1984: U.S. Geological Survey Open-File Report 85–205, 96 pp.

Clayton, G. A., 1983, Geology of the White Pass Area, South-Central Cascade Range, Washington: M.S. thesis, University of Washington, Seattle, 212 pp.

Condie, K. C., and Swenson, D. H., 1973, Compositional variation in three Cascade stratovolcanoes: Jefferson, Rainier, and Shasta: Bulletin Volcanologique, v. 37 (2), p. 205–230.

Conrey, R. M., 1991, Geology and Petrology of the Mt. Jefferson Area, High Cascade Range, Oregon: Ph.D. dissertation, Washington State University, Pullman, 357 pp.

Coombs, H. A., 1936, The geology of Mount Rainier National Park: University of Washington Publications in Geology, v. 3, p. 131–212.

Crandell, D. R., 1963, Surficial geology and geomorphology of the Lake Tapps Quadrangle, Washington: U.S. Geological Survey Professional Paper 388-A, 84 pp.

Crandell, D. R., 1969, Surficial geology of Mount Rainier National Park, Washington: U.S. Geological Survey Bulletin 1288, 41 pp.

Crandell, D. R., 1971, Postglacial lahars from Mount Rainier Volcano, Washington: U.S. Geological Survey Professional Paper 677, 75 pp.

Crandell, D. R., 1973, Potential Hazards from Future Eruptions of Mount Rainier, Washington: U.S. Geological Survey Miscellaneous Investigations Map I-836.

Crandell, D. R., and Miller, R. D., 1974, Quaternary stratigraphy and extent of glaciation in the Mount Rainier region, Washington: U.S. Geological Survey Professional Paper 847, 59 pp.

Crandell, D. R., and Mullineaux, D. R., 1967, Volcanic hazards at Mount Rainier, Washington: U.S. Geological Survey Bulletin 1238, 26 pp.

Crandell, D. R., and Mullineaux, D. R., 1978, Potential hazards from future eruptions of Mount St. Helens Volcano, Washington: U.S. Geological Survey Bulletin 1383-C, 25 pp.

Cullen, J., 1978, Impact of a Major Eruption of Mount Rainier on Public Service Delivery Systems in the Puyallup Valley, Washington: M.S. thesis, University of Washington, Seattle, 202 pp. (Reprinted as Volume 2 of Marts, M. E., Hodge, D. C., Sharp, V. L., Sheridan, F. E., MacGregor, J. M., and Cullen, J. M., 1978, Social Implications of Volcano Hazard: Case Studies in the Washington Cascades and Hawaii: Department of Geography, University of Washington, Seattle, 350 pp.)

Cullen Tanaka, J., 1983, Volcanic Hazards Assessment for Mt. Rainier, Washington: Washington State Department of Emergency Services, Olympia, 28 pp.

DeLoach, S. R., 1989, Global positioning of Mount Rainier—the conclusion: P.O.B (Point of Beginning), v. 14 (3), p. 24–28.

Dillman, D., Schwalbe, M., and Short, J., Jr., 1982, Communication behavior and social impacts following the May 18, 1980 eruption of Mt. St. Helens: in Mt. St. Helens: One Year Later: Cheney, Eastern Washington University Press, p. 173–179.

Driedger, C. L., and Kennard, P. M., 1986, Ice volumes on Cascade volcanoes: Mount Rainier, Mount Hood, Three Sisters, and Mount Shasta: U.S. Geological Survey Professional Paper 1365, 28 pp.

Driedger, C. L., and Walder, J. S., 1991, Recent debris flows at Mount Rainier: U.S. Geological Survey Open-File Report 91–242, 2 pp.

Dzurisin, D., Johnson, D. J., and Symonds, R. B., 1983, Dry tilt network at Mount Rainier, Washington: U.S. Geological Survey Open-File Report 83–277, 9 pp.

Easterbrook, D. J., Briggs, N. D., Westgate, J. A., and Gorton, M. P., 1981, Age of Salmon Springs Glaciation in Washington: Geology, v. 9, p. 87–93.

Easterbrook, D. J., Westgate, J. A., and Naeser, N. D., 1985, Pre-Wisconsin fission-track, paleomagnetic, amino acid, and tephra chronology in the Puget Lowland and Columbia Plateau, Washington [abs.]: Geological Society of America Abstracts with Programs, v. 17, p. 3.

Evarts, R. C., Ashley, R. P., and Smith, J. G., 1987, Geology of the Mount St. Helens area: record of discontinuous volcanic and plutonic activity in the Cascade arc of southern Washington: Journal of Geophysical Research, v. 92, p. 10,155–10,169.

Fink, J. H., 1991, Constructional controls on the degradation of volcanoes [abs.]: EOS, Transactions of the American Geophysical Union, v. 72, p. 227.

Finn, C., and Williams, D. L., 1987, An aeromagnetic study of Mount St. Helens: Journal of Geophysical Research, v. 92, p. 10,194–10,206.

Finn, C., Phillips, W. M., and Williams, D. L., 1991, Gravity Anomaly and Terrain Maps of Washington: U.S. Geological Survey Geophysical Investigations Map GP-988.

Fiske, R. S., 1984, Volcanologists, journalists, and the concerned local public: a tale of two crises in the eastern Caribbean: in Explosive Volcanism: Inception, Evolution, and Hazards: Washington, D.C., National Academy Press, p. 170–176.

Fiske, R. S., Hopson, C. A., and Waters, A. C., 1963, Geology of Mount Rainier National Park, Washington: U.S. Geological Survey Professional Paper 444, 93 pp.

Fiske, R. S., Hopson, C. A., and Waters, A. C., 1964, Geologic Map and Section of Mount Rainier National Park, Washington: U.S. Geological Survey Miscellaneous Geologic Investigations Map I-432.

Frank, D., 1983, Origin, distribution, and rapid removal of hydrothermally formed clay at Mount Baker, Washington: U.S. Geological Survey Professional Paper 1022-B, 31 pp.

Frank, D. G., 1985, Hydrothermal Processes at Mount Rainier: Ph.D. dissertation, University of Washington, Seattle, 195 pp.

Frank, D. G., in press, Surficial extent and conceptual model of hydrothermal system at Mount Rainier, Washington: Journal of Volcanology and Geothermal Research.

Frank, D., Meier, M. F., and Swanson, D. A., 1977, Assessment of increased thermal activity at Mount Baker, Washington, March 1975-March 1976: U.S. Geological Survey Professional Paper 1022-A, 49 pp.

Franklin, J. F., MacMahon, J. A., Swanson, F. J., and Sedell, J. R., 1985, Ecosystem response to the eruption of Mount St. Helens: National Geographic Research (Spring issue), p. 198–216.

Guffanti, M., and Weaver, C. S., 1988, Distribution of late Cenozoic volcanic vents in the Cascade Range: volcanic arc segmentation and regional tectonic considerations: Journal of Geophysical Research, v. 93, p. 6513–6529.

Hadley, K. C., and Lahusen, R. G., 1991, Deployment of an acoustic flow-monitor system and examples of its application at Mount Pinatubo, Philippines [abs.]: EOS, Transactions of the American Geophysical Union, v. 72 (44), p. 67.

Hamilton, R. M., 1992, A perspective on the IDNDR: Natural Hazards Observer (Nov.-Dec. issue), p. 8–9.

Hammond, P. E., and Cole, S. F., 1992, Preliminary investigation of Mount Aix caldera, Cascade Range, southern Washington [abs.]: EOS, Transactions of the American Geophysical Union, v. 73 (43), p. 612.

Hammond, P. E., and Korosec, M. A., 1983, Geochemical analyses, age dates, and flow-volume estimates for Quaternary volcanic rocks, Southern Cascade Mountains, Washington: State of Washington Division of Geology and Earth Resources Open-File Report 83–13, 36 pp.

Harris, S. L., 1988, Fire Mountains of the West: the Cascade and Mono Lake Volcanoes: Missoula, Montana, Mountain Press Publishing Co., 379 pp.

Heaton, T. H., and Hartzell, S. H., 1987, Earthquake hazards on the Cascadia subduction zone: Science, v. 236, p. 162–168.

Hildreth, W., and Fierstein, J., 1983, Mount Adams Volcano and its 30 "parasites" [abs.]: Geological Society of America Abstracts with Programs, v. 15, p. 331.

Hoblitt, R. P., Miller, C. D., and Scott, W. E., 1987, Volcanic hazards with regard to siting nuclear-power plants in the Pacific Northwest: U.S. Geological Survey Open-File Report 87–297, 196 pp.

Iwatsubo, E. Y., and Swanson, D. A., 1992, Trilateration and distance-measuring techniques used at Cascades and other volcanoes: in Ewert, J. W., and Swanson, D. A., editors, Monitoring Volcanoes: Techniques and Strategies Used by the Staff of the Cascades Volcano Observatory, 1980–90: U.S. Geological Survey Bulletin 1966, p. 103–114.

Jacoby, G. C., Williams, P. L., and Buckley, B. M., 1992, Tree ring correlation between prehistoric landslides and abrupt tectonic events in Seattle, Washington: Science, v. 258, p. 1621–1623.

Janda, R. J., Scott, K. M., Nolan, K. M., Martinson, H., 1981, Lahar movement, effects, and deposits: U.S. Geological Survey Professional Paper 1250, p. 461–478.

Johnson, R. E., and Jarvis, J. S., 1980, The Insurance Industry Response: Mount St. Helens Report issued by Washington State Insurance Commission Staff, Olympia.

Kartez, J., 1982, Emergency Planning Implications of Local Governments' Responses to Mt. St. Helens: Working Paper 46, Boulder Institute of Behavioral Science, University of Colorado, Boulder, 29 pp.

Karlin, R. E., and Abella, S. E. B., 1992, Paleoearthquakes in the Puget Sound region recorded in sediments from Lake Washington, U.S.A.: Science, v. 258, p. 1617–1620.

Korosec, M. S., 1989, New K-Ar age dates, geochemistry, and stratigraphic data for the Indian Heaven Quaternary volcanic field, south Cascade Range, Washington: Washington Division of Geology and Earth Resources Open-File Report 89–3, 42 pp.

Lachman, R., and Bonk, W., 1960, Behavior and beliefs during the recent volcanic eruption at Kapoho: Science, v. 133, p. 1405–1409.

Law, L. K., Auld, D. R., and Booker, J. R., 1980, A geomagnetic variation anomaly coincident with the Cascade volcanic belt: Journal of Geophysical Research, v. 85, p. 5297–5302.

Leeman, W. P., Smith, D. R., Hildreth, W., Palacz, Z., and Rogers, N., 1990, Compositional diversity of late Cenozoic basalts in a transect across the southern Washington Cascades: implications for subduction zone magmatism: Journal of Geophysical Research, v. 95, p. 19,561–19,582.

Lees, J. M., 1992, The magma system of Mount St. Helens: non-linear high-resolution P-wave tomography: Journal of Volcanology and Geothermal Research, v. 53, p. 103–116.

Leik, R. K., Leik, S. A., Ekker, K., and Gifford, G., 1982, Under the Threat of Mount St. Helens: a Study of Chronic Family Stress: Family Study Center, University of Minnesota, Minneapolis, 179 pp.

Lipman, P. W., and Mullineaux, D. R., editors, 1981, The 1980 eruptions of Mount St. Helens, Washington: U.S. Geological Survey Professional Paper 1250, 844 pp.

López, D. L., and Williams, S. N., 1993, Catastrophic volcanic collapse: relation to hydrothermal processes: Science, v. 260, p. 1794–1796.

Lowe, D. R., Williams, S. N., Leigh, H., Connor, C. B., Gemmell, J. B., and Stoiber, R. E., 1986, Lahars initiated by the 13 November 1985 eruption of Nevado del Ruiz, Colombia: Nature, v. 324, p. 51–53.

Luetgert, J., Mooney, W. D., Criley, E., Keller, G. R., Gridley, J., Miller, K., Trehu A., Nabelek, J., Smithson, S. B., Humphreys, C., Christensen, N. I., Clowes, R., and Asudeh, I., 1992, Crustal velocity structure of the Pacific NW: the 1991 seismic refraction/wide-angle reflection experiment [abs.]: Geological Society of America Abstracts with Programs, v. 24 (5), p. 66.

Ma, L., 1988, Regional Tectonic Stress in Western Washington from Focal Mechanisms of Crustal and Subcrustal Earthquakes: M.S. thesis, University of Washington, Seattle, 84 pp.

Ma, L., Crosson, R. S., and Ludwin, R., 1991, Focal mechanisms of western Washington earthquakes and their relationship to regional stress: U.S. Geological Survey Open-File Report 91–441–D, p. 1–42.

MacMahon, J. A., 1982, Mount St. Helens revisited: Natural History, v. 91 (5), p. 14–24.

Mader, G., and Blair, M., 1987, Living with a Volcanic Threat; Response to Volcanic Hazards, Long Valley, California: William Spangle and Associates, Portola Valley, California.

Majors, H. M., and McCollum, R. C., 1981, Mount Rainier—the tephra eruption of 1894: Northwest Discovery, v. 2, p. 334–381.

Malone, S. D., and Frank, D., 1975, Increased heat emission from Mount Baker, Washington: EOS, Transactions of the American Geophysical Union, v. 56, p. 679–685.

Malone, S. D., Qamar, A., and Jonientz-Trisler, C., 1991, Recent seismicity studies at Mount Rainier, Washington [abs.]: Seismological Society of America Program, v. 62, p. 25.

Marts, M. E., Hodge, D. C., Sharp, V. L., Sheridan, F. E., MacGregor, J. M., and Cullen, J. M., 1978, Social Implications of Volcano Hazard, Case Studies in the Washington Cascades and Hawaii: Department of Geography, University of Washington, Seattle, 350 pp.

Massonnet, D., Rossi, M., Carmona, C., Adragna, F., Peltzer, G., Feigl, K., and Rabaute, T., 1993, The displacement field of the Landers earthquake mapped by radar interferometry: Nature, v. 364, p. 138–142.

Mattinson, J. M., 1977, Emplacement history of the Tatoosh volcanic-plutonic complex, Washington: ages of zircons: Geological Society of America Bulletin, v. 88, p. 1509–1514.

May, P. J., 1985, Recovering from Catastrophes: Federal Disaster Relief Policy and Politics: Westport, Connecticut, Greenwood Press, 186 pp.

Miller, R. B., 1989, The Mesozoic Rimrock Lake inlier, southern Washington Cascades: implications for the basement to the Columbia Embayment: Geological Society of America Bulletin, v. 101, p. 1289–1305.

Mooney, W. D., and Weaver, C. S., 1989, Regional crustal structure and tectonics of the Pacific coastal states: California, Oregon, and Washington: in Pakiser, L. C., and Mooney, W. D., editors, Geophysical Framework of the Continental United States: Geological Society of America Memoir 172, p. 129–161.

Moxham, R. M., Crandell, D. R., and Marlatt, W. E., 1965, Thermal features at Mount Rainier, Washington, as revealed by infrared surveys: U.S. Geological Survey Professional Paper 525-D, p. 93–100.

Mullineaux, D. R., 1974, Pumice and other pyroclastic deposits in Mount Rainier National Park, Washington: U.S. Geological Survey Bulletin 1326, 83 pp.

Mullineaux, D. R., 1986, Summary of pre-1980 tephra-fall deposits erupted from Mount St. Helens, Washington state, USA: Bulletin of Volcanology, v. 48, p. 17–26.

Newhall, C. G., and Self, S., 1982, The Volcanic Explosivity Index (VEI): an estimate of explosive magnitude for historical volcanism: Journal of Geophysical Research, v. 87, p. 1231–1238.

Norris, R. D., 1991, The Cascade volcanoes: monitoring history and current land management: U.S. Geological Survey Open-File Report 91-31, 68 pp.

Palmer, S. P., Pringle, P. T., and Shulene, J. A., 1991, Analysis of lique-
    fiable soils in Puyallup, Washington: *in* Borcherdt, R. D., and
    Shah, H. C., editors, Proceedings, Fourth International Confer-
    ence on Seismic Zonation, v. 4, p. 621–628.
Perkins, J. B., and Moy, K. K., 1989, Liability for Earthquake Hazards
    or Losses and its Impacts on Washington's Cities and Counties:
    Association of Bay Area Governments, Oakland, California, 4 pp.
Perry, R. W., and Greene, M., 1983, Citizen Response to Volcanic Erup-
    tions: the Case of Mount St. Helens: New York, Irvington Pub-
    lishers, 145 pp.
Perry, R. W., and Lindell, M. K., 1986, Twentieth Century Volcanicity
    at Mt. St. Helens: The Routinization of Life Near an Active Vol-
    cano: Tempe, Arizona State University, 171 pp.
Porter, S. C., 1978, Glacier Peak tephra in the north Cascade Range,
    Washington: stratigraphy, distribution, and relationship to late-
    glacial events: Quaternary Research, v. 10, p. 30–41.
Pringle, P. T., and Palmer, S. P., 1992, Liquefiable volcanic sands in
    Puyallup, Washington correlate with Holocene pyroclastic flow
    and lahar deposits in upper reaches of the Puyallup River valley
    [abs.]: Geological Society of America Abstracts with Programs,
    v. 24, p. 76.
Saarinen, T. F., and Sell J. L., 1985, Warning and Response to the Mount
    St. Helen's eruption: Albany, State University of New York Press,
    240 pp.
Schuster, R. L., Logan, R. L., and Pringle, P. T., 1992, Prehistoric rock
    avalanches in the Olympic Mountains, Washington: Science,
    v. 258, p. 1620–1621.
Scott, K. M., Pringle, P. T., and Vallance, J. W., 1992, Sedimentology,
    behavior, and hazards of debris flows at Mount Rainier, Washing-
    ton: U.S. Geological Survey Open-File Report 90–385, 106 pp.
Sekiya, S., and Kikuchi, Y., 1889, The eruption of Bandai-san: Coll. Sci.
    Imp. Univ. Tokyo, v. 3, p. 91–172.
Sheets, P. D., and Grayson, D. K., editors, 1979, Volcanic Activity and
    Human Ecology: New York, Academic Press, 644 pp.
Sherrod, D. R., and Smith, J. G., 1990, Quaternary extrusion rates of the
    Cascade Range, northwestern United States and southern British
    Columbia: Journal of Geophysical Research, v. 95,
    p. 19,465–19,474.

Siebert, L., 1992, Volcano hazards—threats from debris avalanches: Nature, v. 356, p. 658–659.

Simkin, T., Siebert, L., McClelland, L., Bridge, D., Newhall, C., and Latter, J. H., 1981, Volcanoes of the World: A Regional Directory, Gazetteer, and Chronology of Volcanism During the Last 10,000 Years: Stroudsburg, Pennsylvania, Hutchinson Ross Publishing Co., 232 pp.

Simkin, T., Siebert, L., and McClelland, L., 1984, Volcanoes of the World 1984 Supplement: Washington, D.C., Smithsonian Institution, 33 pp.

Smith, D. R., and Leeman, W. P., 1987, Petrogenesis of Mount St. Helens dacitic magmas: Journal of Geophysical Research, v. 92, p. 10,313–10,334.

Smith, D. R., and Leeman, W. P., 1993, The origin of Mount St. Helens andesite: Journal of Volcanology and Geothermal Research, v. 55, p. 271–303.

Smith, G. O., 1897, The rocks of Mount Rainier: U.S. Geological Survey Annual Report 18, p. 416–423.

Sorensen, J. H., 1981, Emergency response to Mt. St. Helens' eruption: March 20 to April 10, 1980: Working Paper 43, Boulder Institute of Behavioral Science, University of Colorado, Boulder, 70 pp.

Souther, J. G., and Yorath, C. J., 1991, Neogene assemblages: in Gabrielse, H., and Yorath, C. J., editors, Geology of the Cordilleran Orogen in Canada: Geological Survey of Canada, Geology of Canada, no. 4, p. 373–401.

Stanley, W. D., Finn, C., and Plesha, J. L., 1987, Tectonics and conductivity structures in the southern Washington Cascades: Journal of Geophysical Research, v. 92, p. 10,179–10,193.

Stanley, W. D., Gwilliam, W. J., Latham, G., and Westhusing, K., 1992, The southern Washington Cascades conductor—a previously unrecognized thick sedimentary sequence?: American Association of Petroleum Geologists Bulletin, v. 76, p. 1569–1585.

Swanson, D. A., 1992, Geologic map of the McCoy Peak Quadrangle, southern Cascade Range, Washington: U.S. Geological Survey Open-File Report 92-336, 36 pp.

Swanson, D. A., and Evarts, R. C., 1992, Tertiary magmatism and tectonism in an E-W transect across the Cascade arc in southern Washington [abs.]: Geological Society of America Abstracts with Programs, v. 24 (5), p. 84.

Swanson, D. A., Brown, J. C., Anderson, J. L., Bentley, R. D., Byerly, G. R., Gardner, J. N., and Wright, T. L., 1979, Preliminary structure contour maps on the top of the Grande Ronde and Wanapum Basalts, eastern Washington and northern Idaho: U.S. Geological Survey Open-File Report 79-1364.

Swanson, D. A., Malone, S. D., and Samora, B. A., 1992, Mount Rainier: a Decade Volcano: EOS, Transactions of the American Geophysical Union, v. 73, p. 177, 185-186.

Swanson, D., Malone, S., and Casadevall, T., 1993, Mitigating the hazards of Mount Rainier: EOS, Transactions of the American Geophysical Union, v. 74 (12), p. 133.

Taylor, E. M., 1990, Volcanic history and tectonic development of the central High Cascade Range, Oregon: Journal of Geophysical Research, v. 95, p. 19,611-19,622.

Tolan, T. L., and Beeson, M. H., 1984, Exploring the Neogene history of the Columbia River: discussion and geologic field trip guide to the Columbia River Gorge, Part 2, Road log and comments: Oregon Geology, v. 46, p. 103-112.

University of Washington Geophysics Program, 1993, Western Washington Earthquake Catalog 1969-1992: University of Washington, Seattle.

Vance, J. A., Clayton, G. A., Mattinson, J. M., and Naeser, C. W., 1987, Early and middle Cenozoic stratigraphy of the Mount Rainier-Tieton River area, southern Washington Cascades: Washington Division of Geology and Earth Resources Bulletin 77, p. 269-290.

Voight, B., Glicken, H., Janda, R. J., and Douglass, P. M., 1981, Catastrophic rockslide avalanche of May 18: U.S. Geological Survey Professional Paper 1250, p. 347-377.

Walsh, T. J., Korosec, M. A., Phillips, W. M., Logan, R. L., and Schasse, H. W., 1987, Geologic Map of Washington—Southwest Quadrant: Washington Division of Geology and Earth Resources Geologic Map GM-34.

Warrick, R. A., Anderson, J., Downing, T., Lyons, J., and Ressler, J., 1981, Four Communities Under Ash: After Mount St. Helens: Monograph Number 34, University of Colorado, Institute of Behavioral Science, Boulder, 150 pp.

Waters, A. C., 1961, Keechelus problem, Cascade Mountains, Washington: Northwest Science, v. 35, p. 39–57.

Weaver, C. S., 1976, Seismic Events on Cascade Volcanoes: Ph.D. dissertation, University of Washington, Seattle, 189 pp.

Weaver, C. S., and Smith, S. W., 1983, Regional tectonic and earthquake hazard implications of a crustal fault zone in southwestern Washington: Journal of Geophysical Research, v. 88, p. 10,371–10,383.

Williams, D. L., and Finn, C., 1987, Evidence for a shallow pluton beneath the Goat Rocks Wilderness, Washington, from gravity and magnetic data: Journal of Geophysical Research, v. 92, p. 4867–4880.

Williams, D. L., Abrams, G., Finn, C., Dzurisin, D., Johnson, D. J., and Denlinger, R., 1987, Evidence from gravity data for an intrusive complex beneath Mount St. Helens: Journal of Geophysical Research, v. 92, p. 10,207–10,222.

Wood, C. A., and Kienle, J., 1990, Volcanoes of North America: Cambridge, U.K., Cambridge University Press, 354 pp.

Yamaguchi, D. K., 1983, New tree-ring dates for recent eruptions of Mount St. Helens: Quaternary Research, v. 20, p. 246–250.

Yamaguchi, D. K., 1985, Tree-ring evidence for a two-year interval between recent prehistoric explosive eruptions of Mount St. Helens: Geology, v. 13, p. 554–557.

APPENDIX A
GROWTH MANAGEMENT ACT OF
STATE OF WASHINGTON

The following is a brief review of the State of Washington's Growth Management Act (GMA) of 1990, its requirements, how those requirements are to be implemented, how policies and regulations concerning the control of development in volcanic hazard zones will be effected, and some of the potential problems with respect to volcanic hazards and growth in Pierce County, which together with King County, would bear the brunt of most hazardous events at Mount Rainier.

On April 1, 1990, the State of Washington passed the GMA, requiring those counties and municipalities meeting certain population trends to develop comprehensive growth management plans. Under the act, unincorporated Pierce County is required to adopt a comprehensive growth management plan dealing with nine elements (Environment and Critical Areas, Land Use, Rural Areas, Housing, Transportation, Utilities, Capital Facilities, Economic Development, and Community Plans) by July 1, 1993, and to have regulations in place implementing the plan by July 1, 1994. In addition, the 18 municipalities within Pierce County are required to produce their own growth management plans. All 19 plans and development regulations affecting Pierce County must be consistent, so that the town of Orting, for example, does not treat a growth management issue differently from the way unincorporated Pierce County, or the city of Puyallup, treats it.

The GMA requires public participation in the design of the comprehensive plan. The act specifies that a Citizen's Advisory Group (CAG), consisting of residents representing various interests identified in the GMA, be appointed by the elected county executive. In Pierce County, the

CAG consists of 31 appointed citizens who represent interests ranging from agriculture to utilities.

The CAG's responsibility is to recommend policies to the County Council that will be used as guidelines for writing the regulations and ordinances to implement the growth management plan. The CAG's policy recommendations are based on issues identified by the GMA, by the county government, by Pierce County citizens in a countywide survey, and in public workshops conducted by the CAG and by members of Advisory Committees on Elements (ACEs).

For unincorporated Pierce County there are nine ACEs, each dealing with one of the nine elements. The ACEs are chaired by a CAG member, and their membership consists of citizens who are concerned with various issues. The CAG makes final policy recommendations for approval by the Pierce County Council. At any step in this process, changes in recommended policy can occur.

One of the responsibilities of the ACE on Environment and Critical Areas is recommending growth management policies to deal with volcanic hazards in unincorporated Pierce County. The ACE members who considered volcanic hazard policies consisted of approximately 20 individuals, including two professional geologists. The Environment and Critical Areas ACE developed the following working guidelines with respect to geologic hazards:

> Residents of Pierce County live in areas where they are exposed to various natural hazards which to varying degrees endanger lives and properties. The primary objective of the Environment and Critical Areas ACE with respect to these natural hazards is to recommend growth management policies which minimize this risk to the lives, property, and resources of the citizens of Pierce County by: 1) using the best available data and methodologies to identify, evaluate, and delineate hazardous areas; 2) informing present and potential property owners as to the existence and nature of the specific natural hazards which endanger their lives and property through written disclosure prior to

property sale, through deed and plat notification, and through public education programs; 3) directing development away from areas subject to catastrophic, life-threatening natural hazards where the hazard cannot be mitigated; 4) requiring appropriate standards for site development and for the design of structures in areas subject to natural hazards where the effects of such hazards can be mitigated; 5) establishing land use practices in hazardous areas where development does not cause or exacerbate natural processes which endanger the lives, property, and resources of the citizens of Pierce County.

The best available data at the time that the ACE considered the issue of volcanic hazards (July 1992) was from Crandell and Mullineaux (1967) and Crandell (1971, 1973). The new study by Scott and others (1992) on debris flows from Mount Rainier was not available. In the course of this assessment, the ACE recognized the need for more and better data on the nature, frequency, extent, and areas effected by volcanic events at Mount Rainier. The publications by Crandell and Mullineaux clearly show that the most significant volcanic hazard confronting the residents of Pierce County are debris flows in the Puyallup, Carbon, White, and Nisqually River valleys. Although the Crandell-Mullineaux papers lack adequate information on recurrence intervals and probabilities for various-sized debris flows, the ACE agreed that the maximum event for planning consideration was a debris flow similar in size to the Electron-debris flow, as described in Crandell (1971). After much debate and discussion the ACE unanimously recommended the following:

The only significant volcanic hazard affecting Pierce Co. are lahars and glacier runs from Mount Rainier. The volcanic hazard area, based on the work of Crandell (1971), Crandell & Mullineaux (1967), Crandell (1973, USGS Map I-836, Potential Hazards, Mount Rainier Washington), consists of the area rated at moderate and high risk on USGS Map I-836 and of the Carbon River

Valley based on work in progress by Pringle (Washington DNR), Scott (USGS) & Vallance (USGS).

Committee Recommendations (unanimous approval):

1) Prohibit the construction of critical facilities in the volcanic hazards area.

2) Require notification by the county of the volcanic hazards for property situated in the volcanic hazards area on all property deeds and recorded plats.

3) Establish an educational program to educate the citizens of Pierce County as to volcanic hazards.

4) Discourage all uses of the land in the volcanic hazards area except agriculture and recreation.

5) Prohibit any further expansion of public facilities in the volcanic hazards area (trunk lines going through the volcanic hazards area are allowed).

6) Explore creative incentives/alternatives for public acquisition of property in volcanic hazard areas (transfer of development rights).

7) Refer discussion of warning systems and evacuation plans to the Pierce County Emergency Management Department.

Upon submitting these recommendations to the CAG (September 14, 1992), several CAG members were concerned about the word "Prohibit" in recommendations 1 and 5 and asked if the ACE did not mean "Limit." The CAG was assured that the "Prohibit" was intended and was to be applied to areas in which the probability for inundation by a debris flow during a 100-year period was 9.5 percent or greater (Scott and others, 1992). This issue remains unresolved by the CAG and the Pierce County Council as of April 1993.

Another problem concerns existing development and future needs in the upper Nisqually Valley in the Ashford-Elbe areas, where the probability for inundation by a debris flow is substantially greater than 9.5

percent in a 100-year period. As a gateway to Mount Rainier National Park, the upper Nisqually Valley is the most intensively used outdoor recreational area in Pierce County. The hundreds of thousands of tourists visiting the Ashford-Elbe areas sorely tax the capacity of the communities to fill their needs. The CAG needs to resolve the dilemma of serving the needs of these visitors—for example rest areas serviced by sewer systems—while avoiding intense development such as that at Yosemite in an area which could be inundated by a debris flow with less than a half-hour warning. A warning system such as flow detectors or seismometers telemetered to sirens in Ashford and Elbe and an emergency evacuation plan are clearly needed.

Because the GMA requires consistency among 19 comprehensive plans for Pierce County (unincorporated Pierce County and 18 municipalities within it), additional major problems can be easily anticipated. For example, the town of Orting, which lies within the volcanic hazard zone defined above, plans expansion. Orting is responsible for preparing its own comprehensive plan. There may be great difficulty in arriving at consistency between the unincorporated Pierce County and Orting comprehensive plans.

Pierce County is expected to grow from its present population of 580,000 to about 740,000 by the year 2010. To say that there is intense economic pressure to provide homes, schools, and other needed facilities for many of these people by developing the volcanic hazard area in Pierce County is an understatement. Individuals with financial interests in seeing that this development occurs are well represented on the CAG and Pierce County Council. Owners of property within the volcanic hazard zone have no reluctance in exercising their political and legal rights in protecting their financial interests. There will be strong pressure from various groups to substantially weaken the recommendations cited above.

APPENDIX B
SCIENTIFIC RESEARCH AND COLLECTING IN MOUNT RAINIER NATIONAL PARK

Following is a document prepared by representatives of Mount Rainier National Park that outlines procedures for conducting research in the park.

Scientific research has long been an important part of the operation of national parks. Park Service Management Policies direct that a program of natural and social science research be conducted in the parks to support National Park Service goals, and to assist park staff in carrying out the mission of the Service by providing accurate scientific information for planning, development, and management of the parks.

The National Park Service cooperates with research institutions, and in recognition of the scientific value of parks as natural laboratories, investigators are encouraged to use the parks for scientific studies, when such use is consistent with National Park Service policies. Research activities that might disturb resources or visitors, that require the waiver of any regulation, or that involve the collection of specimens, are allowed only pursuant to the terms and conditions of an appropriate permit. Manipulative or destructive research activities generally are not permitted in national parks. Exceptions may be granted if the impacts will be short lived, the park is the only area where such research can be conducted, the value of the research is greater than the resource impacts, or the research is essential to provide information for resource management.

Scientific collecting activities that involve the removal of plants, animals, minerals, or archeological, historical, or paleontological objects are allowed only if they are (1) proposed in conjunction with authorized research activities, and (2) authorized and conducted in accordance with all applicable legislation, regulations, and guidelines.

Scientific collecting in national parks is specifically controlled by the regulations published in the Code of Federal Regulations, Title 36, Chapter 1, Section 2.5. These regulations, as they apply to Mount Rainier National Park, allow us to issue permits for the collection of geologic specimens only. Permits may be issued only to an official representative of a reputable scientific or educational institution or a State or Federal agency for the purpose of research, baseline inventory, monitoring, impact analysis, group study, or museum display when the park superintendent determines that the collection is necessary to the stated scientific or resource management goals of the institution or agency. A permit cannot be issued if removal of the specimen would result in damage to other natural or cultural resources or adversely affect environmental or scenic values, or if the specimen is readily available outside of the park.

At Mount Rainier National Park, collection permits are issued only after the park superintendent approves a written research proposal and determines that the collection will benefit science or has the potential for improving the management and protection of park resources. When permission is granted to conduct research and/or collect specimens, there are several additional requirements that must be met. First, all specimens retained in displays or collections must be labeled with official National Park Service labels. Second, specimens must be entered into the National Park Service National Catalog. Third, these specimens and the data derived from them must be made available to the public, and copies of all reports and publications resulting from a research specimen collection permit will be filed with the park superintendent. We place these reports in the park library, where they are available to the public.

If you wish to conduct scientific research within Mount Rainier National Park, you must prepare a written research proposal, and present it to the park superintendent for review. This proposal should present evidence that the work will provide new and/or useful information. If possible it should demonstrate how the information may be useful for park management and resource protection. Methods of collection should be presented in detail so that the impact of collecting activity on other park resources and park visitors can be accurately determined. You may address proposals to:

> Park Superintendent
> Mount Rainier National Park
> Tahoma Woods
> Star Route
> Ashford, WA 98304

When your proposal has been reviewed and approved, a permit can be issued for specimen collection and research. If specimens are to be retained, official National Park Service labels and catalog work sheets will be provided for your use. In October of each year that research continues, you will be sent a short Investigator's Annual Report, to be completed and returned by December 15. This report will provide information for a National Park Service published report on research in the national parks.

APPENDIX C
WORKSHOP ATTENDEES

James L. Anderson	*Univ. Hawaii, Hilo*
Charles R. Bacon	*USGS, Menlo Park*
George Bergantz	*Univ. Washington*
David D. Blackwell	*Southern Methodist Univ.*
Russell Blong	*Macquarie Univ., Australia*
Steven Brantley	*USGS, Vancouver*
Sutikno Bronto	*Merapi Volcano Observatory, Indonesia*
Ernie Campbell	*Boeing Company*
Rebecca Carmody	*USGS, Reston*
Gerardo Carrasco-Núñez	*UNAM, Mexico*
Tom Casadevall	*USGS, Denver*
Kathy Cashman	*Univ. Oregon*
John R. Delaney	*Univ. Washington*
Carolyn L. Driedger	*USGS, Vancouver*
Mike Dungan	*Southern Methodist Univ.*
John Dvorak	*USGS, Vancouver*
Al Eggers	*Univ. Puget Sound*
Michael J. Ellett	*Univ. Washington*
John F. Ferguson	*Univ. Texas, Dallas*
Carol Finn	*USGS, Denver*
Richard S. Fiske	*Smithsonian Institution*
Duncan Foley	*Pacific Lutheran Univ.*
David Frank	*EPA, Seattle*
Peter Frenzen	*U.S. Forest Service, Amboy, Washington*
Bruce Hanshaw	*National Research Council*
Cathie Hickson	*Geological Survey of Canada, Vancouver*
Peter R. Hooper	*Washington State Univ.*
D. R. Johnson	*National Park Service, Seattle*

Deborah S. Kelley	*Univ. Washington*
Stephen C. Kuehn	*Washington State Univ.*
Jonathan Lees	*Yale Univ.*
Tommy Leland	*Univ. Washington*
Peter Lipman	*USGS, Menlo Park*
Jim Luetgert	*USGS, Menlo Park*
Gary Machlis	*Univ. Idaho*
Jon Major	*USGS, Vancouver*
Steve Malone	*Univ. Washington*
Larry G. Mastin	*USGS, Vancouver*
Otoniel Matias	*INSIVUMEH, Guatemala*
Peter May	*Univ. Washington*
I. S. McCallum	*Univ. Washington*
Juliet McKenna	*Univ. Washington*
Richard B. Moore	*USGS, Denver*
Seth C. Moran	*Univ. Washington*
Mark T. Murphy	*Pacific Northwest Laboratory*
Thomas M. Murray	*Boeing Commercial Airplane Group*
Chris Newhall	*USGS, Reston*
Arnold Okamura	*USGS, Hawaii National Park*
Kenneth H. Olsen	*Lynnwood, Washington*
Paul Kilho Park	*NOAA, Washington, D.C.*
Patrick Pringle	*Washington State Department of Natural Resources*
Anthony Qamar	*Univ. Washington*
Emmanuel Ramos	*Philippine Institute of Volcanology and Seismology*
Vincent J. Realmuto	*JPL*
Mark Reid	*USGS; Menlo Park*
Mike Rhodes	*Univ. Massachusetts*
Malcolm Rutherford	*Brown Univ.*
Barbara Samora	*Mount Rainier National Park*
J. Eric Schuster	*Washington State Department of Natural Resources*
Tom Sisson	*Univ. Texas, Dallas*

Dal Stanley	*USGS, Denver*
André Stepankowsky	*Longview, Washington,* Daily News
Trileigh Stroh	*Univ. Washington*
Donald A. Swanson	*USGS, Seattle*
Darin Swinney	*Mount Rainier National Park*
Tom Usselman	*National Research Council*
James W. Vallance	*McGill Univ.*
Dina Venezky	*Brown Univ.*
Barry Voight	*Pennsylvania State Univ.*
Richard B. Waitt	*USGS, Vancouver*
Courtenay Wilkerson	*Univ. Washington*
Colin J. N. Wilson	*Univ. Cambridge*
Edward W. Wolfe	*USGS, Vancouver*
Kirby Young	*Pennsylvania State Univ.*
David R. Zimbelman	*USGS, Denver*